T0259708

SpringerBriefs in Energy

More information about this series at http://www.springer.com/series/8903

Hossein Seifi · Hamed Delkhosh

Model Validation for Power System Frequency Analysis

 Springer

Hossein Seifi
Faculty of Electrical and Computer
 Engineering, Iran Power System
 Engineering Research Centre (IPSERC)
Tarbiat Modares University (TMU)
Tehran, Iran

Hamed Delkhosh
Faculty of Electrical and Computer
 Engineering
Tarbiat Modares University (TMU)
Tehran, Iran

ISSN 2191-5520   ISSN 2191-5539   (electronic)
SpringerBriefs in Energy
ISBN 978-981-13-2979-1   ISBN 978-981-13-2980-7   (eBook)
https://doi.org/10.1007/978-981-13-2980-7

Library of Congress Control Number: 2018958945

This Springer imprint is published by the registered company Springer Nature Singapore Pte Ltd.
The registered company address is: 152 Beach Road, #21-01/04 Gateway East, Singapore 189721,
Singapore

Preface

The power industry and its associated interconnected electric grid are, perhaps, the most important infrastructure of the civilized world. The electric grid operation is highly challenging from various viewpoints, requiring detailed studies, typically carried out using simulation tools available for such studies. Modelling is the major step of any type of simulation studies.

Modelling for power system frequency analysis is addressed in this research monograph. The framework, methods, and formulations are given based on a practical research as carried out by Iran Power System Engineering Research Centre (IPSERC) for the Iran Grid Management Company (IGMC), implemented on the large-scale Iranian power grid (15th in the world, in terms of the installed generation capacity, roughly 78 GW-2018).

We intend to bridge the gap between formal learning of the model validation process from the frequency perspective and its practical implementation for a real and large-scale power grid. Therefore, the intended audiences of this monograph are both power industry experts and academia.

Some individuals and organizations have made the writing of this book possible. We should especially thank the experts in IGMC, including Dr. M. Rajabi Mashhadi, Mr. A. Mazkoori, Mr. E. Zabihzadeh, and Mr. A. Abbasszadeh, for their valuable discussions and comments. The monograph has been reviewed by our team members on IPSERC, namely Mrs. M. Sajadi, Mr. M. Jorjani, Mr. V. Hakimian, Mr. L. Heidari, Mr. M. Yousefian, and Mr. S. Tajrobekar, who provided us useful suggestions and comments. Sincere thanks are due to Dr. D. Merkle, Dr. R. Premnath, and their colleagues, from Springer, for their support in the preparation and publication of this monograph. Last but not least, we should thank our families who accepted us as part-time family members during this monograph preparation.

Finally, we should mention that although we have attempted to review the monograph in order to be error-free, some might still exist. Please feel free to email us about possible errors, comments, opinions, or any other useful information.

Tehran, Iran   Hossein Seifi
October 2018   Hamed Delkhosh

Contents

Chapter 1
Introduction

Abstract This chapter provides an introduction to power system frequency control (primary, secondary, and tertiary control levels) and briefly explains different issues and challenges in this regard. Moreover, this chapter aims to clarify the important role of model validation in the planning and the operation studies of a power system, which illustrates the significance of system-wide model validation from the frequency point of view. Moreover, the structure of the subsequent chapters is also presented.

Keywords Frequency control · Model validation · Primary control
Secondary control · Tertiary control · Governor modelling
Automatic Generation Control (AGC)
Under Frequency Load Shedding (UFLS) · Ancillary services

1.1 Frequency Control

A power system is the largest or one of the largest systems ever made by humankind. This highly interconnected system may span over a country, some countries or even a continent. In recent decades, this system has encountered numerous planning and operational challenges, which are progressively becoming more complicated due to various technological and non-technological issues. Of these challenges, power system stability is an issue of great concern in both the planning and operational phases.

With the emerge of new technologies in power systems (such as renewable energies, distributed generations, micro-grids, smart grids, etc.) and new control devices in generation as well as non-generation (transmission, sub-transmission and distribution) levels, and also higher integration of power grids, power system stability has received more attention from the industry viewpoint. If this challenges are not paid attention properly and studied in various phases, it may deteriorate system performance significantly.

Power system stability studies are divided into various types such as rotor angle, transient, voltage, and frequency stability. Performing simulation from each perspective needs taking into account particular considerations and increasing the accuracy of specific parts of modelling.

Power system frequency control is categorized into primary, secondary and tertiary control levels. The primary control reserve is provided in the first few seconds following an active power imbalance mainly through inertial response and governor action of the generating units. The first goal of primary frequency control is to prevent the frequency from entering the Under Frequency Load Shedding (UFLS) and over-speed generator relays activation area and its second goal is to stabilize the frequency within an acceptable secure range.

The activated primary frequency reserve is maintained until it is replaced by the secondary control reserve either manually by the system operator or automatically which is known as Automatic Generation Control (AGC). By the secondary control, the frequency is often recovered to its nominal value and the power interchanges between areas are set to the specified values. Then, the tertiary control reserve comes into play, which adjusts the active power production of the generating units often according to some economic calculations and releases the activated secondary control reserve.

It is worth noting that in interconnected power systems with multi areas of operation, the secondary and tertiary reserves should be provided by the area which is the source of incident whereas the primary reserve is provided by all of the generating units connected to the synchronous area.

The primary control can be considered mainly as a technical problem whereas the secondary control is usually considered as a technical/economic problem. Therefore, proper economic compensation should be observed for the generating units involved in AGC. It is worth noting that even in primary level, the proper financial motivation is necessary for having enough participation of players. Meanwhile, the tertiary level is almost merely an economic problem.

Among these three control levels, the primary control has received the least attention especially in highly developed power systems, mainly because of the following reasons:

- Primary frequency has been considered mandatory for all traditional generating units usually without any economic compensation,
- The operating reserves of the units have sufficiently been high so that in response to a typical incident of the grid, the units are capable of maintaining and recovering the frequency to a stable and secure value,
- Highly integrated grids would guarantee a robust performance of system primary frequency response due to the availability of numerous generating units for most contingencies.

On the other hand, in less developed power systems, the primary control is still a challenging technical issue, as:

- The reserve may not be sufficiently high,
- Not all of the generating units participate in primary control because the penalties/rewards are not significant,
- Some practical issues such as improper tuning of the affecting parameters may limit the response of the generating units so that the overall primary control performance of the system is not acceptable.

Nevertheless, due to increasing penetration level of low inertia distributed generation (such as most of the wind power generating units) from one hand and developments in various electric market issues (such as payments for ancillary services) on the other hand, the primary control is being paid more attention for both highly developed and less developed power systems. Therefore, it is now an issue of R&D in which the economic aspects should also be observed and at the same time, the technical issues should be resolved for different types of power systems.

1.2 Model Validation

Power system planning and operation require simulation studies, based on proper models with valid parameters values, in order to properly investigate the system performance in response to an incident. Both optimistic and pessimistic models have crucial drawbacks that make realistic models essential for secure and economic power system operation and planning. It is worth noting that utilization of valid models for studies is a more vital issue in stability limited systems.

The frequency stability studies is a system-wide problem (see Sect. 3.2) as the frequency itself is a global parameter. Therefore, all of the affecting components should be properly modelled and considered in order to study and analyse this type of the problem. Among the different components, the modelling of the turbine-governor of the generating units is of great importance when studying from the power system frequency point of view, especially the primary level.

It should be noted that, unfortunately, the models and their respective parameters are not often readily or fully accessible. This issue is even more challenging in large-scale systems, involving hundreds of generating units participating in frequency control. Although some concepts such as equivalencing can be employed to facilitate the modelling process, it results in omitting some components such as generating units, which have to be observed for enhancing system frequency performance.

A Supervisory Control And Data Acquisition (SCADA) system in the dispatching centre of a real power system provides valuable data of system frequency, as well as generating units responses, following the incidents. On the other hand, it is a question of developing a framework in which, instead of using the detailed modelling for each unit, some standard models are employed with the identified affecting parameters (from the frequency perspective) so that based on available SCADA information, the frequency behaviour of the system can be validated. By using this idea, detailed modelling of each generating unit may be avoided.

While this subject, namely *Model Validation for Power System Frequency Analysis*, should be performed for a specific system, the framework, methods, and formulation can be employed for different power systems. Moreover, implementing this process for a real and large power grid comprises lots of practical issues which some of them are addressed in this monograph.

1.3   Monograph Structure

This monograph outlines a model validation framework from the power system frequency perspective and its implementation for the real and large-scale Iranian power grid, hereafter, called the *test grid*, in order to show the procedure of implementation.

Chapter 2 (*Fundamentals of Frequency Control*) is devoted to basic principles of frequency control with the following topics:

- Physical phenomenon,
- Basic principles,
- A historical view,
- Frequency control requirements,
- Parameters affecting system frequency behaviour,
- Frequency control strategies: Principles,
- Frequency control strategies: Europe and North America.

In Chap. 3 (*Theoretical Aspects of Model Validation from Frequency Perspective*), algorithms, methods and formulation of model validation from the frequency point of view are discussed with the following structure:

- Introduction,
- Model validation approaches,
- Power system frequency model validation: practical experiences,
- Generalized model validation framework,
- Initial measures,
- Static data correction method,
- Quantitative validation,
- Frequency and voltage dependency of loads.

Numerical results of implementing the framework for the *test grid* are presented in Chap. 4 (*Implementation and Numerical Results*), with the following structure:

- The test grid,
- Initial measures,
- First approach: existing models,
- Second approach: new generic model.

Chapter 5 (*Detailed Sensitivity Analysis*) investigates the sensitivity of the validation performance with respect to various parameters and measures with the following structure:

- Introduction,
- Basic parameters and measures,
- Scenarios,
- Simulation results.

Chapter 2
Fundamentals of Frequency Control

Abstract In this chapter, the concepts regarding frequency, as a basic parameter of power systems are described. Initially, the physical phenomenon, based on that, this parameter is affected and how it can be controlled are explained. A short historical view on frequency is then followed. It is then discussed how the players in a power system may be affected by frequency. A more detailed description of the affecting parameters on power system frequency related to the generating units (primary frequency participation, droop, dead band, practical powers, operating point, activity range, frequency ramp rate, time constants, and inertia) and loads (frequency and voltage dependency), is then given. The final sections are devoted to various frequency control strategies especially from UCTE and ENTSO-E (both for Europe) and NERC (North America) perspectives, as the most famous governing entities for two large-scale interconnected power systems.

Keywords Frequency control strategies · Operating reserves
Frequency control requirements · Frequency history · Governor · Inertia · Droop
Dead band · Ramp rate · Load modelling

2.1 Physical Phenomenon

Electricity plays an important role in the everyday life of human beings. A consumer, either large or small, considers the supplied electricity of being good, if:

- It is supplied without interruption,
- Its quality is acceptable.

The first point is obvious. Regarding the second point, a consumer typically looks at the quality in terms of acceptable supplied voltage. While voltage is a critical parameter, another important parameter is frequency. It is easy to check the nameplate of an electric appliance to see its operating frequency (mostly 50 or 60 Hz), as well as its operating voltage.

Voltage issue is not in the scope of this monograph. Frequency is the parameter to be discussed. A power system is considered to be healthy if its frequency is kept in a

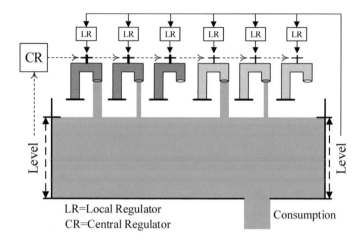

Fig. 2.1 A simple physical analogy for power system frequency and its controls

narrow range around the nominal value (say 50 or 60 Hz). However, various factors would affect the frequency. Therefore, power system engineers have developed both manual and automatic emergency actions to keep it within an acceptable range.

A simple physical analogy for power system frequency and its controls is depicted in Fig. 2.1. Consider a water pool for which there is a consumption as well as different types of generation, controlled by their valve positions. Six taps are considered (from the left) as follows:

1. Fully open, slow valve position change,
2. Partially open, slow valve position change,
3. Out of service, can be called on upon requirement, slow valve position change,
4. Fully open, fast valve position change,
5. Partially open, fast valve position change,
6. Out of service, can be called on upon requirement, fast valve position change.

The consumption varies continuously. The aim is to keep the level fixed or within acceptable small range. For each tap, there is one Local Regulator (LR) which continuously checks the level and modifies its valve position (if practically possible and just for in-service taps) a bit so that if the level increases, the valve closes and vice versa. Assume the capability of tap position change is limited to say, 5% of its original nominal position.

If the consumption of the pool is increased significantly, although all LRs would try to keep the level fixed, they may be unsuccessful to do the job. It is now the time that a Central Regulator (CR) comes into play. This central regulator may be controlled either automatically or manually by the operator. It somehow changes the nominal positions of the valves and checks the level to be kept fixed.

As some taps which were already partially open, may become fully open now, the taps manoeuvre capabilities may become limited for any later level change. That is why some new taps may be put in service or out of service in case of consumption decrease. If done, the positions of the taps of the former stage (those that became fully open from partially open or became fully close in case of consumption decrease) can be brought back to their nominal values.

Let's call these three stages as follows:

- Local Regulators (LRs) acting on all taps: *primary control*,
- Manual or Automatic Central Regulator (MCR or ACR) acting on some taps to change their valves nominal positions: *secondary control*,
- Switching over some parts of the regulation process to new taps (with possible modifications of the valve positions of existing taps): *tertiary control*.

On the other hand, the consumption change can be of different types:

(a) Gradual with a slow rate,
(b) Gradual with a high rate,
(c) Sudden and large,
(d) Sudden and small.

For any type of mentioned consumption change, the three types of control (namely primary, secondary, and tertiary) are activated to keep the level fixed. It is obvious that the primary control is fast acting, followed by the secondary and then the tertiary controls. It should be noted that similar type (gradual/sudden, large/small, and with a slow/high rate) of tap valve position has the same kind of effect on water level but obviously in the reverse direction. It means that consumption increase (decrease) or generation decrease (increase) has the same effect.

Now, let's look at the power system and its resemblances with the described pool. A power system is similar to a water pool with the water replaced by electricity. The consumption is replaced by the electricity loads distributed among different buses. The water inflows of taps are replaced by electricity productions of the generating units. Similar to water level, the power system frequency is used as the basic control parameter. In terms of control activities, the followings apply:

- Each generating unit is equipped with a local controller, namely governor (primary control, activated typically in the range of 0–30 s),
- Some generating units are controlled, remotely, by a central controller, either manually or automatically [Automatic Generation Control (AGC)]. This controller is normally located in the main dispatching centre of a power system and acts through some types of communication networks from the central controller to the generating units (vice versa) (secondary control, activated typically in the range of 30 s–15 min),
- There is typically a central decision maker, either in or close to the main dispatching centre, which monitors the secondary control activities. This decision maker decides to alter the generation levels of the units or commit/de-commit units, often considering some economic indices. Normally, these types of decisions are activated manually (tertiary control, activated typically in the range of some minutes to an hour),

Fig. 2.2 Schematic of
supplying a load through
mechanical and inertial
powers

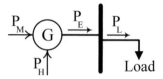

- Similar to the water pool, generation and load changes of the power system can be of different types,
- Similar to the taps, generating units production can be changed at different rates.

Considering these explanations, the basic steps of this research monograph for executing the model validation for the *test grid* are as follows:

(a) The measured frequency changes over a period of 0–120 s are selected for studies (scope of the primary and secondary control levels),

(b) The secondary control activities are determined through available data in the central dispatching centre (i.e., how the power generation of the units have changed, either manually or automatically, from 30 s onward),

(c) The speed governors of the generating units are modelled based on available information (models and parameters values),

(d) The system frequency is simulated for 0–120 s, considering the activities in step (b) and the models as assumed in step (c),

(e) By comparing the simulated frequency response in step (d) and the measured frequency of step (a), the models as used in step (c) are improved and validated, using techniques and methodologies as described in Chap. 3.

2.2 Basic Principles

Consider a simple case (Fig. 2.2) in which a generating unit (G) supplies a fixed load (P_L, assumed to be independent of voltage and frequency[1]). Hence:

$$P_E = P_L \tag{2.1}$$

where P_E is the electrical power supplied by both the mechanical input power to the turbine (P_M) and the inertial power of the generator (P_H). Therefore:

$$P_E = P_M + P_H \tag{2.2}$$

Provided $P_E = P_M$, $P_H = 0$, and the generator speed would be constant. If this balance is disturbed for any reason, (either by changing P_M and/or P_E), P_H changes and the generator accelerates or decelerates, according to:

---

[1] See Sect. 2.5.2.

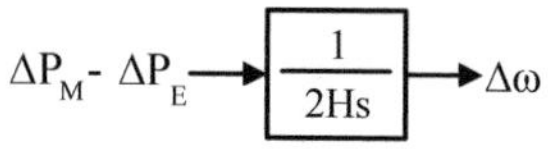

Fig. 2.3 The block diagram of frequency dynamic behaviour

$$2H\frac{d\Delta\omega}{dt} = \Delta P_M - \Delta P_E \tag{2.3}$$

where H is the inertial constant of the generator, in seconds, and the other parameters are in per unit. The block diagram of this equation is depicted in Fig. 2.3.

In practice, there are a large number of generating units (N) in a power system. If H_i and f_i are respectively the inertial time constant and the frequency of the generating unit i, the frequency of the system, as a global (system-wide) parameter, is affected by H_i and f_i of all the units as follows:

$$f_{COI} = \frac{\sum_{i=1}^{N} H_i \times f_i}{\sum_{i=1}^{N} H_i} \tag{2.4}$$

where f_{COI} is called Centre Of Inertia (COI) frequency.

2.3 A Historical View

Once the Westinghouse Electric (WE) decided to install a generating unit in Niagara, 25 Hz was chosen as the nominal frequency due to the technological limitations (number of poles, inertia, etc.), dictated at that time. While 25 Hz was also used in Boston by Hartford Company for an electric locomotive, it was not considered as a standard value and some other frequencies were also employed world-wide (like 12.5 Hz in Europe). The main problem with low frequency electric power generation was the flickers, quite visible, with the electric discharge lamps. That is why initially the generating units based on 40 Hz and later on based on 50 and 60 Hz were designed, constructed and utilized.

In the late 19th century, WE decided to expand its activities for a large power grid. While its engineers recommended the use of 133 Hz as the nominal frequency, Tesla proposed 60 Hz as the most appropriate choice, based on various types of analysis and tests. His proposal was accepted and chosen by WE as a standard value in the USA. Although 220 V was initially chosen by Tesla as the nominal voltage, later on, this value was changed to 110 V, mainly due to safety reasons.

At more or less the same time, AEG in Germany decided to develop its equipment based on 110 V and 50 Hz. After WWII, the voltage in Europe was upgraded to 220 V.

Also Great Britain changed its nominal frequency from 60 to 50 Hz so that this value was selected as the nominal standard frequency, throughout Europe.

While the nominal voltage and frequency are typically the same throughout a country, different values are used in various parts of some countries. Brazil (110–127 and 220 V) and Japan (50 and 60 Hz) are two typical cases. It is worth noting that although the power system efficiency by using 60 Hz (in comparison with 50 Hz) is higher,[2] 50 Hz is more common throughout the world.

2.4 Frequency Control Requirements

Frequency control requirements and issues can be investigated from different viewpoints as follows:

(a) *From generating unit viewpoint*: The turbine-generator may be subjected to severe mechanical damage of the shaft and turbine blade due to possible mechanical resonances. Moreover, the V/Hz[3] operational region of an industrial generator[4] should be limited to a narrow band which controls the thermal losses of the generator,

(b) *From transmission equipment viewpoint*: Similar to the generators, the transmission equipment performances are also adversely affected by the substantial change in the system frequency. Moreover, the system elements impedances (reactance and capacitance) are frequency-dependent. This may lead to some operational problems,

(c) *From load viewpoint*: The frequency is important from the load point of view for two reasons:

   – **Continuity of supply**: If frequency deviation is large, the loads may have to be curtailed (either manually or automatically through UFLS[5] relays) to keep system frequency within an acceptable range,

   – **Quality of supply**: Any substantial change in frequency may lead to overheating of electric appliances. Moreover, the speed and the output power of most conventional motors are proportional to frequency so that their responses may be adversely affected.[6] Speed drive motors responses are, however, independent of system frequency. While some clocks may also operate based on

---

[2]Due to reasons such as efficiency and also increase of material usage in production process of motors, generators and transmission equipment.

[3]Proportional to magnetic flux.

[4]IEC 34-1 standard.

[5]Under Frequency Load Shedding (UFLS).

[6]Any frequency increase normally leads to increase in power of the motors. This is beneficial for the power system as it can result in controlling the frequency (see Sect. 2.5.2).

system frequency, current digital clocks behaviours are independent of system frequency.

(d) *From transmission operator viewpoint*: Based on what described in (a), (b), and (c), the transmission system operator should try to keep the frequency within an acceptable range by available resources. The operator should do so in response to various incidents and different types of load change. The resources should be properly allocated and the required reserves should be distributed among them with consideration of all the operational limits.[7]

Next section deals with the parameters affecting the system frequency behaviour. These parameters are considered for the validation process, in this monograph.

2.5 Parameters Affecting System Frequency Behaviour

The parameters of the two main components of the power system, namely generating units and loads would affect the system frequency behaviour, as detailed in the following sections.

2.5.1 Generating Units

As discussed in Sect. 2.2, any mismatch between the mechanical input to and the electrical output from the generating unit results in a change in accelerating power and accordingly the generator speed. In order to reduce the mismatch, the primary controller of a generating unit, namely governor (like local regulator), detects the system frequency variation (like water level change) and changes the valve/gate position of the turbine (like the tap position) and accordingly the power generation (like water inflow). The way it behaves is shown in Fig. 2.4, where:

$$R = \frac{\Delta f^{p.u.}}{\Delta P_M^{p.u.}} = \frac{\Delta f/f_n}{\Delta P_M/P_n} \tag{2.5}$$

In which R is the droop parameter and f_n and P_n are the nominal frequency and power, respectively. P_{gen} is the active power of generating unit in the nominal frequency, which will be called the operating point, hereon. As shown in Fig. 2.4, if the frequency drops, the governor acts in a way to increase the mechanical power of the unit (by opening the valve/gate of the turbine), thereby, stabilize the rotor speed and hence the system frequency. R is dimensionless and is in the range of 2–12% (typically 4–5%).[8] Although the governor of a unit tries to control the frequency,

---

[7]The methodologies for allocation and distribution may be economic, technical, etc.

[8]Although the actual slope is negative, R is usually presented as a positive value.

Fig. 2.4 Droop
characteristics of a governor

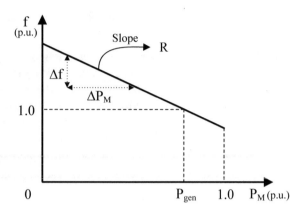

Table 2.1 Droop effect on the unit participation in compensating a total power imbalance

Scenario	First	Second
Droop	$R_1 = 0.04$, $R_2 = 0.04$	$R_1 = 0.03$, $R_2 = 0.05$
Unit participation	$\Delta P_1 = 8\,\text{MW}$, $\Delta P_2 = 8\,\text{MW}$	$\Delta P_1 = 10\,\text{MW}$, $\Delta P_2 = 6\,\text{MW}$
Frequency deviation (unit 1)	$-0.04 = \frac{\Delta f_1/50}{8/100} \rightarrow \Delta f_1 = -0.16\,\text{Hz}$	$-0.03 = \frac{\Delta f_1/50}{10/100} \rightarrow \Delta f_1 = -0.15\text{Hz}$
Frequency deviation (unit 2)	$-0.04 = \frac{\Delta f_2/50}{8/100} \rightarrow \Delta f_2 = -0.16\text{Hz}$	$-0.05 = \frac{\Delta f_2/50}{6/100} \rightarrow \Delta f_2 = -0.15\text{Hz}$

it cannot restore the frequency to its nominal value and is just successful in its stabilizing by increasing the mechanical power.

The performance of the governors of the generating units in a multi-machine case can be explained by a two-unit example, as follows:

Assume two similar units with the droop parameters R_1 and R_2 are supplying a fixed[9] 160 MW load, each with 80 MW generation. Assume the load is increased by 10% (16 MW). The following equations apply:

$$R_1 = \frac{\Delta f/f_n}{\Delta P_1/P_{n1}}, \quad R_2 = \frac{\Delta f/f_n}{\Delta P_2/P_{n2}}, \quad \Delta P_1 + \Delta P_2 = \Delta L \qquad (2.6)$$

The above three equations can be solved for ΔP_1, ΔP_2 and Δf. The results of two different scenarios are shown in Table 2.1.

As demonstrated:

• Based on its droop parameter, each generating unit takes over some part of the load increment so that if its R is lower, unit participation is higher and vice versa,
• Frequency is stabilized at a new level, as affected by the droop parameters.

Although the droop characteristic is shown as in Fig. 2.4, the actual behaviour is somewhat different due to a parameter defined as Dead Band (DB). This parameter

---

[9]Independent of frequency.

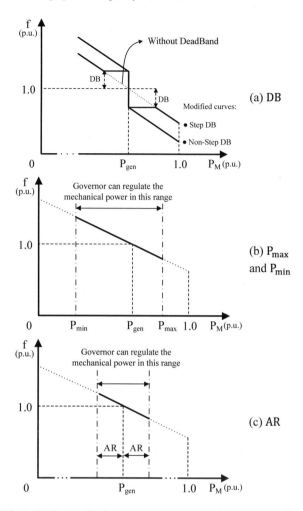

Fig. 2.5 The Effect of different affecting parameters on droop characteristics of a governor

is introduced to reduce the mechanical stresses on valve/gate in such a way that the governor would respond only if $|\Delta f| \geq$ DB.

Two types of this intentional DB may be employed:

- Step DB, in which $|\Delta f|$ is passed through when $|\Delta f| \geq$ DB,
- Non-step DB, in which $|\Delta f| -$ DB is passed through when $|\Delta f| \geq$ DB.

These two are shown in Fig. 2.5a. To avoid higher stresses, especially around the DB region, the non-step DB is normally used. A typical value for DB is 16.7 mHz.[10]

Besides R and DB, the third parameter to be observed for a generating unit is its generation capability. In fact, two such values, P_{max} and P_{min} are defined as the

---

[10]Current value for North America, while 36 MHz was earlier used.

maximum and minimum generation capacity of a unit. While P_{max} may be highly affected by the ambient conditions[11] (0.8 per unit of its nominal value, as an example), P_{min} is limited by thermodynamics instability of a thermal unit (0.4 per unit of its nominal value, as an example). Between these two values is the practical range in which mechanical power can be set as shown in Fig. 2.5b.[12] Therefore, P_{max} and P_{min} should be considered for the later process of the system frequency behaviour.

Besides R, DB, P_{max} and P_{min}, another parameter of interest is the unit Activity Range (AR), which shows the limitations imposed on the deviation of mechanical power (P_M) from its operating point (P_{gen}). AR (say 0.1 per unit) is usually logically dictated by the governor of a generating unit as shown in Fig. 2.5c.

Following the description of R, DB, P_{max}, P_{min}, and AR, the Frequency Ramp Rate (FRR) should be mentioned as an important parameter affecting the frequency behaviour. FRR is logically considered in the governor system to limit the thermal stresses on the mechanical components of the unit. While there is a ramp rate limitation on the opening/closing of a turbine valve for its protection, FRR[13] is normally much lower than this physical ramp rate (0.2 p.u./s). For a hydro turbine, such logical limitation is not normally imposed as no thermal gradient happens there.

Besides the parameters discussed so far, there are various time constants for the turbine and governor block diagrams which also affect the system frequency behaviour. Moreover, a generating may not participate in primary frequency control due to temporary/permanent technical problems or not having enough financial motivation. This can be modelled by a logical binary parameter, namely Primary Frequency Participation (PFP). Also, the relation between the frequency rate of change after a power imbalance with the inertia (H) was described in Sect. 2.2.

So, briefly speaking, the following parameters should be considered in term of the generating units:

- Primary Frequency Participation (PFP),
- Dead Band (DB),
- Droop (R),
- Activity Range (AR),
- Operating point (P_{gen}),
- Frequency Ramp Rate (FRR),
- Maximum and minimum generation capabilities (P_{max} and P_{min}),
- Time constants (T),
- Inertial constant (H).

A schematic diagram of a turbine-governor model, considering these affecting parameters, is presented in Fig. 2.6.[14]

---

[11] For instance, a gas turbine generation power is more sensitive to the ambient conditions.

[12] While a single continuous acceptable range is normally defined for a thermal unit, there may be some acceptable ranges and forbidden zones for a hydro unit.

[13] For a 160 MW gas turbine, typical normal and fast values are 11 and 30 MW/min, respectively.

[14] Inertial constant (H) is usually modelled in generator.

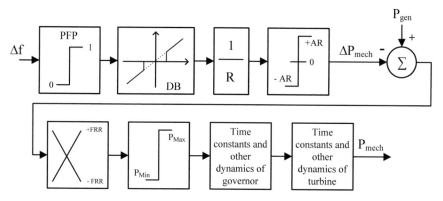

Fig. 2.6 A schematic of a turbine-governor model, considering affecting parameters

The mechanical output behaviour of a turbine in response to a frequency decrement is shown in Fig. 2.7, for some typical cases. As shown in Fig. 2.7a,[15] a lower value of R results in a higher mechanical power change for identical frequency deviation [see Eq. (2.5)]. Figure 2.7b shows a delayed response with a higher frequency deviation required with increasing DB. As in Fig. 2.7c is demonstrated, whenever the power generation of the unit reaches its maximum capability (P_{max}), the mechanical power is limited at that value. For a specific change in frequency, a higher mechanical power change can happen with a higher FRR (see Fig. 2.7d).

2.5.2   Loads

The load dependency on voltage and frequency is an important issue of concern for the validation process. Although on voltage and/or frequency variations, a load changes dynamically, it is of common practice to consider its static behaviour as follows[16]:

$$P = P_0 \times \left(\overline{V}\right)^a \times \left(1 + K_{pf}\Delta f\right) \tag{2.7}$$

$$Q = Q_0 \times \left(\overline{V}\right)^b \times \left(1 + K_{qf}\Delta f\right) \tag{2.8}$$

where P_0 and Q_0 are the nominal active and reactive powers, respectively (i.e. at 1.0 p.u. voltage and nominal frequency). "a" ("b") is the parameter that shows the active (reactive) power dependency on voltage and K_{pf} (K_{qf}) shows the active (reactive) power dependency on frequency. $\overline{V}$ is the load voltage in p.u.

---

[15]In this figure, it is assumed that DB $= 0$ Hz.
[16]See also [43].

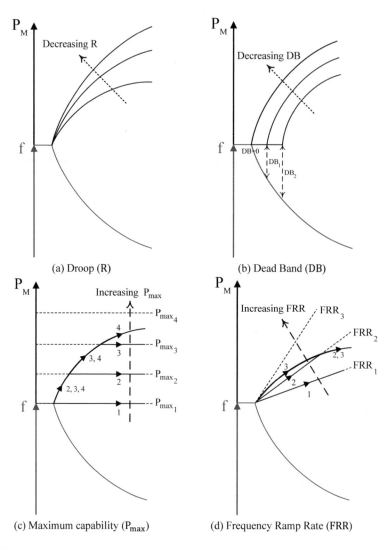

Fig. 2.7 The effect of different affecting parameters on mechanical power

As shown in (2.7) and (2.8), a nonlinear load dependency on voltage is assumed, while due to quite low variation of frequency, a linear load dependency is assumed on the frequency. Following any change in voltage and/or frequency, both P and Q change which have to be observed for the validation process.

The parameters as in (2.7) and (2.8) (i.e. "a", "b", K_{pf}, and K_{qf}) can be calculated for a load point of a system, either through a component-based approach or through a measurement methodology which are out of this monograph scope.

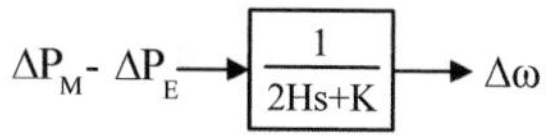

$$\Delta P_M - \Delta P_E \longrightarrow \boxed{\dfrac{1}{2Hs+K}} \longrightarrow \Delta\omega$$

Fig. 2.8 Block diagram of frequency dynamic behaviour considering load dependency on frequency

Parameter "a" is in the range of 0.5–1.5 (typically 1.0), while parameter b is in the range of 1.5–6.0 (typically 2.0). K_{pf} is in the range of 0–3.0 (typically 1.5) and K_{qf} is in the range of −2 to 0 (typically 0).

As reactive power is less sensitive to frequency, for the validation process, the following models are assumed:

$$P = P_0 \times (\overline{V})^a \times (1 + K \times \Delta f) \tag{2.9}$$

$$Q = Q_0 \times \overline{V}^2 \tag{2.10}$$

where "a" and "K"[17] should somehow be determined during the validation process, in case that accurate data is not available. (2.3) can now be stated as follows:

$$2H\frac{d\Delta\omega}{dt} + K\Delta\omega = \Delta P_M - \Delta P_E \tag{2.11}$$

Therefore, Fig. 2.3 is updated as Fig. 2.8.

2.6   Frequency Control Strategies: Principles

As noted earlier, any imbalance between the input and output powers of a grid would result in some change in frequency. Such imbalance can be of different types and origins as follows:

(a) *Step-type*, in which a major and sudden change in active power balance would happen due to a generating unit trip out or a large load interruption,
(b) *Ramp-type*, in which the imbalance would happen in a ramp condition due to:

– Loading of a generating unit (predictable),
– The gradual change of a large industrial load (unpredictable),
– Severe environmental condition change in a region which leads to change of renewable generations output power (unpredictable).

---

[17]"K" is sometimes called *self-regulation coefficient of the load* or *load damping*, as once frequency is reduced, the load would also be reduced, thereby, helping to regulate the frequency.

It should be mentioned that the ramp-type imbalances are categorized as being:

– *Slow* (say 4% of the system load in 30 min),
– *Fast* (say 2% of the system load in 5 min).

(c) *Fluctuating-type*, in which the imbalance would happen due to, say, a small change of domestic loads and/or generation of renewable resources.

The basic tools available for the system operator to regulate the frequency, in response to various mentioned imbalances, are the different types of operating reserves (spinning/non-spinning, manual/automatic, etc.). The reserves are, typically, the generations, potentially available from the resources so that if required, can be utilized. However:

• They are distributed throughout the power system,
• Their characteristics are different.

This is the task of the system operator to observe the mentioned points for the allocation and selection of reserves so that upon the occurrence of any type of imbalance, the frequency can be appropriately regulated.

Generally, following any type of mismatch, the strategy is to:

(a) Engage all or nearly all the generating units to regulate the frequency mainly through governor response (primary control) (0–30 s),
(b) Prepare the system to cope with the next possible imbalance. It means that the primary reserves as used up in (a), should be brought back. This implies that some generating units as advocated for this stage should be rescheduled, either automatically or manually (secondary control) (30 s–15 min),
(c) Consider releasing the secondary reserves as used up in (b), by re-scheduling or committing new generating units (tertiary control) (15 min–1 h). This is typically done based on some technical but mainly non-technical (say economic) consideration.

2.7 Frequency Control Strategies: Europe and North America[18]

2.7.1 Europe

UCTE[19] and ENTSO-E[20] are the two entities within Europe, with the former established in 1999 and the later took over the responsibility from 2009. Both have some types of policies regarding frequency control strategies.

[18]See the References.
[19]Union for the Coordination of the Transmission of Electricity (UCTE).
[20]European Network of Transmission System Operators for Electricity (ENTSO-E).

UCTE has developed some strategies for frequency control which are still in practice by some of the European transmission operators. In UCTE, primary, secondary, tertiary and time control levels are defined, as below:

- Primary control, activates within seconds in order to compensate for the active power imbalance,
- Secondary control, replaces the primary control, so as to be ready for subsequent imbalances, and brings back the frequency to its nominal value after minutes,
- Tertiary control, frees secondary control by re-scheduling generation so as to supply the power grid load in an economic manner,
- Time control, corrects time deviations of the synchronous time in the long term.

It should be noted that primary control and time control are performed as a joint action of all undertakings/TSOs, whereas secondary control and tertiary control reserves should be supplied by the responsible undertakings/TSOs only.

In ENTSO-E, the following strategies are defined:

- Frequency containment process (same as primary control), stabilizes the post-incident frequency at a steady-state value within an acceptable frequency deviation range by activating the Frequency Containment Reserve (FCR) within the whole synchronous area,
- Frequency restoration process (same as secondary control), restores the frequency to its nominal value and replaces the activated FCR by activating the Frequency Restoration Reserve (FRR[21]) in the disturbed area,
- Reserve replacement process (same as tertiary control), restores the activated FRR and/or supports the FRR by activating the Replacement Reserve (RR) in the disturbed area. For GB[22] and IRE[23], the reserve replacement process replaces FCR and FRR.

2.7.2 North America

All parts of the U.S.A and some parts of Mexico and Canada obey the rules and standards as developed by NERC.[24] The defined steps for frequency control in NERC are the same as UCTE (primary, secondary, and tertiary control).

During the early stages of NERC establishment, the primary frequency control received less attention, as all the generating units were equipped with governors which were able to provide an acceptable primary frequency performance in response to incidents.

---

[21]The abbreviation is the same as the frequency ramp rate, which may be misleading.

[22]Great Britain.

[23]Ireland.

[24]North American Electric Reliability Corporation (NERC).

In the early 90s, EPRI[25] conducted a research to show that this performance is degrading over the time. However, the situation was not considered to be critical at that time. Following the introduction of ancillary services in electric markets, EPRI conducted a study on providing various types of such services regarding frequency performance. Some specific technical and non-technical reports as performed by EPRI in the early 21st century provided the basis for the establishment of frequency-based ancillary services.[26]

However, the secondary control received attention, right from the establishment of NERC, as it was mainly handled by the system operators. These evolved over the time:

- Development of Control Performance Criteria (CPC)—A1/A2 and B1/B2,
- Development of Control Performance Standard (CPS) to replace CPC A1/A2,
- Development of Disturbance Control Standard (DCS) to replace CPC B1/B2,
- Development of Balancing Authority ACE[27] Limit (BAAL) as a complement to CPS.

Moreover, the following standards are developed which are directly related to the frequency:

- BAL-001: Real Power Balancing Control Performance,
- BAL-002: Disturbance Control Performance,
- BAL-003: Frequency Response and Frequency Bias Setting,
- BAL-004: Time Error Correction,
- BAL-005: Automatic Generation Control,
- BAL-006: Inadvertent Interchange.

More information regarding frequency control strategies and different issues about frequency control all over the world are addressed in references. Next chapter presents the theoretical aspects of model validation from the frequency perspective.

---

[25] Electric Power Research Institute (EPRI).

[26] See the references.

[27] Area Control Error (ACE).

Chapter 3
Theoretical Aspects of Model Validation from Frequency Perspective

Abstract In this chapter, the theoretical aspects of a framework for power system model validation from frequency viewpoint, including formulations and methods, are presented and described. This framework is applicable to large power systems and uses a quantitative method for comparing the measured and the simulated responses. In order to reduce the unnecessary complexities and efforts, the proposed framework is based on two main approaches and for each, uses the maximum capability of turbine-governor models. Since the framework considers the manual actions of the system operators and other emergency actions after an incident, the time frame of the validation can be chosen arbitrarily based on the availability of data. It should be mentioned that this framework also uses a method to increase the accuracy of the static data before an incident. The frequency and voltage dependency of loads are also observed in the framework even if accurate data is not available.

Keywords Model validation · System-wide modelling · Power system modelling Generating unit modelling · Governor modelling · Load voltage dependency Load frequency dependency · Generalized framework · Static data correction Quantitative matching indices

3.1 Introduction

In various stages of power system operation, planning and control, some types of studies have to be developed and conducted. These studies are basically static or dynamic in nature. The static type refers to the studies describing the steady-state behaviour of the system while the dynamic type refers to the studies for non steady-state conditions. The governing equations used for the former studies are typically algebraic equations, while for the latter, they are differential equations; being either linear or nonlinear.

A typical static type is load flow in which the steady-state conditions of a power system is studied. For a dynamic case, stability (the rotor angle, transient, voltage, and frequency stability) studies are the typical cases. For both types, the components which should be modelled, have to be identified properly. As an example, in a load flow study, the steady-state characteristics of the generating units, loads and transmission elements should be modelled appropriately. For a short-term stability study (typically covering a time frame up to 30 s), the models (differential equations or block diagrams) associated with some elements such as synchronous generators, excitation systems, governors, loads, etc. have to be considered.

It is evident that more reliable results are achieved from such studies in which the used models and the employed parameters are more accurate. Unfortunately, the situation is not as simple as it initially seems in a real power system, as:

- The models are not available for many components,
- Even if the models are available, the parameters as used in these models may not be available,
- Even if the models and the parameters are available, the parameters may have been changed over the time, say, due to ageing,
- There are a tremendous number of elements in a large-scale system, which make the life difficult if someone tries to develop such appropriate models and parameters for all elements.

Fortunately, power system behaviour is monitored continuously through various measuring devices installed in most parts of the system. Some valuable information of various parts are available including—but not limited to—the national control centre, the regional control centres, the power plant control centres, etc. An important use of such information is for the *model validation* in which by comparing the measured performance of the system with what developed by simulation (either static or dynamic), the used models and the parameters studies may be checked and corrected if required.

This chapter is devoted to such an important issue from the power system frequency viewpoint. Initially, various model validation approaches are described. Thereafter, some experiences regarding the validation for power system frequency analysis are briefly reviewed. Following that, a framework to do such a task for a large-scale power system is developed and described. It is also discussed that how the input data for such a framework should be properly generated and used.

3.2 Model Validation Approaches

The model validation approaches may be generally categorized as:

- *Component-based*,
- *System-wide*,
- *Hybrid*.

As the name refers, the first category deals with the validation process for a component or an equipment, such as a generator, an excitation system, etc. Based on some types of tests on the component and comparison with the simulated process, the associated model can be validated. This approach sometimes requires the equipment to be isolated from the system for the testing process.

The component-based approach is itself for:

- *Generation elements* (such as the generator, excitation system, turbine, governor, distributed generation, etc.),
- *Transmission elements* (such as transmission line, transformer, phase shifting transformer, SVC,[1] FACTS[2] devices, etc.),
- *Loads.*

In a system-wide validation approach, the system-wide response is employed in terms of one or some specific variables and the models of the components are considered as a whole. For instance, system frequency is a system-wide variable. However, as the associated important variables, the mechanical outputs of the turbines have to be considered, too. In a system-wide approach, the aim is to validate the models used throughout the system in such a way that the system frequency as well as the mechanical output powers of the generating units would match together. We may not bother what would happen to other variables. Such an approach may be employed for some already available monitored incidents. Once validated, the developed and validated models may be used for analysing the system performance in response to some foreseen incidents in future. While the basics are simple, the practical application of a system-wide approach necessitates a large task of collecting a huge amount of proper information after the incident (such as relay operations, operator manual actions, etc.); otherwise, the validation process would be poor. It also necessitates good engineering judgments throughout the process. These topics are summarized in this monograph.

In a hybrid approach, the detailed model is developed for the under consideration part of the system, while the rest is modelled by its equivalent. Such equivalent model may be developed through the measurements available in the main part boundaries with the rest of the system, say by PMUs.[3]

In terms of the algorithms in matching the measured performance and the simulated response, model validation may be categorized into:

- *Trial and error*, in which based on some types of sensitivity analysis and known parameters ranges, the user would try to validate the model. It is obvious that this task puts a large burden on the user, as there are numerous parameters to be varied, checked and analysed,
- *Mathematical*, in which some types of mathematical algorithms such as Kalman filter is employed for the validation process (involving two stages, namely prediction and correction),
- *Meta-heuristic*, in which some types of algorithms, such as Genetic Algorithm (GA), Particle Swarm Optimization (PSO), etc., are used for such purpose.

---

[1] Static VAR Compensator (SVC).

[2] Flexible Alternating Current Transmission System (FACTS).

[3] Phasor Measurement Unit (PMU).

Next section deals with some practical experiences for model validation in power system frequency analysis. They are only a few such studies and while providing good insight into the problem, they suffer some drawbacks in terms of both the validation process and applying to large-scale power systems. A general framework would then developed in Sect. 3.4.

3.3 Power System Frequency Model Validation: Practical Experiences[4]

There are some published studies regarding model validation for power system frequency analysis. In this section, some are reviewed to make the reader familiar with some relevant concepts while some drawbacks or limitations would be also observed. Later on, we would describe a detailed framework for such validation.

(a) *United Kingdom.* This study is based on a reduced 21-node model of the UK power system. Load flow studies are used to relate generation and load conditions of the nodes. Typical dynamic models and parameters are employed for the reduced system. Model validation is based on the comparison of the maximum frequency deviation for the measured and the simulated cases. PSS/E is used as the simulation software.

As described, the study is based on a reduced model for the base conditions, so new topology configurations cannot be studied. Moreover, it is not possible to validate the model in terms of the mechanical output generations of the units, as there is no such detailed modelling in the approach.

(b) *Brazil.* The study as detailed for Brazil compares the simulation results with those of measured conditions from two perspectives:

- *Qualitative.* By a general comparison between the results, including the stability behaviour, variables (such as P, Q, V, and f), etc.,
- *Quantitative.* By comparison of the damping as well as oscillation frequencies of the simulated and measured cases.

The study has not attempted to validate the models by their modifications and/or the parameters. It provides some hints on why the measured and the simulated results are different, including:

- Low precision of the pre-incident condition,
- Unavailability of load models,
- Imprecise measurements,
- Unavailability of the emergency actions following the incident.

(c) *Cyprus.* Asea Brown Boveri (ABB) conducted a similar study for the small-scale Cyprus grid with 1100 MW generation capacity. For the validation purpose,

---

[4]See the References.

some modifications are done such as changing the generation limits of the units, inertial constants, governors droop and adding dead bands to the governor loops. The loads are considered to be constant current for the active part and constant impedance for the reactive part. The load frequency coefficient is considered to be fixed (1%). It is qualitatively claimed that the results show good matching performances.

(d) *WECC.*[5] An extensive research on a large-scale system is carried out in WECC. It is basically based on the following three stages:

 - *Development*, a basic case is developed using one reference incident,
 - *Validation*, the models are validated through the data available from the generating unit owners and some other incidents,
 - *Verification*, the models as validated above are verified by comparing the simulated and the measured performances for some other incidents.

It is justified that by using a generic governor model (GGOV1), and by tuning of just one control parameter, good results can be achieved, even if, some basic parameters (such as droop and time constants) are modelled while some (such as dead band) are not.

It is indicated that different types of generating units responses (base loaded, responsive without a controller, and responsive with slow or fast controller) can be modelled by such a generic model. Moreover, it is demonstrated that the matching is qualitatively good. However, the approach as described cannot be generalized to all types of power grids.

(e) *Eastern Interconnection (EI)*. In this study, initially, the measured performance of the system frequency is compared with what resulting from the simulation; using the available governors models. The discrepancies observed are shown to be attributed mainly to dead band parameter, which is not properly modelled. Therefore, a generic model, namely WSIEG1, is selected in which dead band is modelled. It is then shown how the parameters of the existing models (namely GAST, IEEEG1, IEEESGO, and TGOV1) can be migrated to the parameters of WSIEG1.

The validity of the used methodology is demonstrated through results. Again, a quantitative analysis is not provided. Moreover, for the specific system under consideration, frequency variation is quite small even for some large incidents. Therefore, the effect of even a small dead band is quite noticeable. In practice, the situation may not be the same for all power grids so that some other affecting parameters may not be detected if the same methodology is used for other grids.

To overcome the problems as discussed so far, a general framework is proposed in the next section so that a robust validation process is achieved for a large-scale system in which:

---

[5]Western Electricity Coordinating Council.

- There are various types of generating units (including thermal, hydro, etc.),
- Accurate and validated governor models are not available for all units, but there are some common types of models and parameters available for some units,
- Accurate load models are not available for the buses,
- Enough information regarding various system variables are available for pre-incident conditions (using SCADA or some types of monitoring systems),
- There is good information available for the emergency actions involved after an incident (either manual or automatic).

3.4   Generalized Model Validation Framework

In this section, a framework is described for the model validation, which can be mainly used for power system frequency analysis. Its basic features are as follows:

- The matching indices between the measured and the simulated responses are made quantitative,
- The matching process is performed for several incidents of different natures, to make the results more reliable,
- All capabilities of available models and data are initially used. A step by step procedure is then followed for modification of the models and the parameters to improve the validation process,
- A longer time frame (0–120 s) is utilized for the validation process in comparison with what normally used (0–30 s). This would improve the validation process,
- A methodology is employed to provide reliable information regarding the pre-incident situation. This would also substantially improve the validation process,
- Load dependency on voltage and frequency are considered for active parts of the loads. The relevant coefficients are assumed unavailable, as it is the case in the most of power grids. These are changed for a known possible range, to find the best matching of the measured and the simulated responses.

The framework is shown in Fig. 3.1. Some explanations are as follows:

(a)   The framework is defined and examined for different types of approaches (parameter i) and incidents (parameter j),
(b)   Some initial measures are carried out as follows (for details, see Sect. 3.5):

- Providing data (as much as possible) of pre-incident conditions,
- Correcting the above data for the instant of the incident (see Sect. 3.6),
- Gathering the dynamic models and parameters, as available,
- Selecting proper test incidents for the study,
- Gathering the information required regarding the actions carried out (either manually or automatically) during and after the incident, using SCADA,
- Adjusting the pre-incident frequency to match with the measured value.

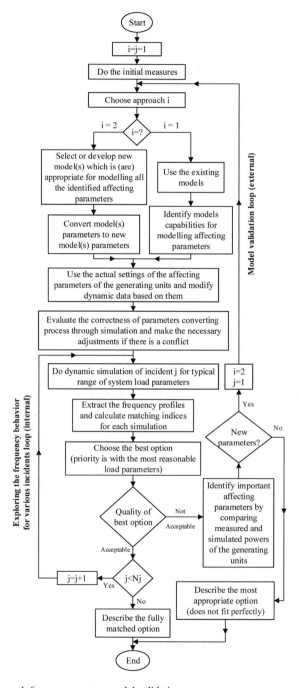

Fig. 3.1 Framework for power system model validation

(c) The procedure of the framework for the validation purpose is as below:

– Initially, with all existing models and parameters, the framework proceeds with the validation process in an external loop. If the matching indices (see Sect. 3.7) are good enough, the approach would be continued for checking a good matching performance for some other incidents, as well. This is shown as the internal loop,

– If the above step is unsatisfactory due to lack of good parameters and/or models, a modified approach has to be tried (i = 2) in which by comparing the measured variations of system variables (especially the active power variations of the generating units) with those of the simulation studies, the affecting parameters are identified first. Then considering the identified parameters, new generic models should be selected or user-defined models should be developed. In this step:

  • The expanded model is employed in such a way that its parameters can be easily extracted from the one already available,
  • It is initially checked that the performance of the system with these new models are the same as what is already available if we ignore the new elements of the new models. This step is required to ensure that the conversion has been done correctly,
  • Due to the different nature of the incidents, new parameters may have to be considered in the model. Therefore, if the simulations are satisfactory for a specific incident, it has to be tested for some other incidents, too (going through the internal loop),
  • User-defined models, as may be required in the above step, are tried to be developed based on existing standard models. If again unsuccessful, new models should be tried.

– The above-mentioned steps are implemented with the best knowledge of the dependency of loads on voltage and frequency. Different values for these coefficients are also tried for better matching (see Sect. 3.8),

– If the whole story fails, the validation process would be terminated and noted as being unsuccessful.

It should be mentioned that in all steps, besides the system frequency behaviour, the active powers of the generating units may also be used for the comparison purposes. If the framework is followed properly, normally good results are achieved. Some reasons causing model validation failure are:

• Introduction of strict matching indices,
• Specific loads, which cannot be properly modelled by "a" and "K" coefficients,
• Substantial errors in various system parameters (such as resistances, reactances, time constants, etc.),
• Unavailability of reliable information regarding pre and post-incident conditions,
• Unsuccessful identification of all affecting parameters on system frequency.

Following sections provide more details on various stages of the framework.

3.5 Initial Measures

The following steps are carried out in the *initial measures* block:

(a) Gathering the static and dynamic available data (such as the generators models and parameters, different time constants, transmission lines and transformers static parameters, etc.),

(b) Selecting the test incidents. In this selection, various types of small and large incidents should be selected with both types of frequency increment and decrement. Moreover, all pre and post-incident information, as well as the closest snapshot (a file including all data of the grid for a specific instant) to the incident instant should be available through the monitoring system,

(c) Taking the snapshot, closest to the instant of the incident for each test, from the state estimator, as the initial pre-incident database. We would see how we can modify it considering the power system conditions at exactly the instant of the incident,

(d) Gathering the information available before, during, and after each incident, through the SCADA system. This valuable information would be used in various stages of the validation process. The system frequency and the active power generations of the generating units are of special concern,

(e) Correction and modification of the database as provided in step (c), to provide a modified database for the instant of the incident (for details, see Sect. 3.6). This would significantly improve the validation process. It mainly concentrates on correction of ON and OFF status of the generating units, as well as, their operating points (P_{gen}) and practical maximum and minimum active power generations (P_{min}, P_{max}); through what gathered in step (d),

(f) Identification of the details of the incidents, through what gathered in step (d), including some items such as the type and size of each incident and the emergency actions taken during and after each of them,

(g) Correction of the initial (pre-incident) frequency of the simulations, to match with the measured value. This is needed, as it highly affects the quality of the validation process, due to nonlinear behaviour of the droop characteristics (because of the presence of the dead band parameter in the governor loop). Such correction can be done by a slow ramp-type increase or decrease of the system load in the simulation stage,

(h) Specification of the ranges for "a" and "K" for load modelling. As discussed earlier (see Sect. 2.5.2), the load dependency on voltage and frequency plays important role in the validation process and is considered in the framework by changing the relevant coefficients for some specified ranges. Some details would be provided in Sect. 3.8.

Except (e) and (h) for which some details are provided below, the details of the other steps would become more clear, once we provide the numerical results in Chap. 4.

3.6   Static Data Correction Method

A method for static data (i.e. steady state condition of the generating units before the incident occurrence) correction is presented in this section. This method, shown in Fig. 3.2, modifies the ON and OFF status, the operating point (P_{gen}), and the maximum and minimum practical powers (P_{max}, P_{min}) of the generating units according to available information from SCADA. It is worth mentioning that in this method 0, SCADA, and * superscripts indicate the extracted values from the initial static snapshot, extracted values from the SCADA information, and the revised output values of each parameter, respectively.

Initially, the information of each generating unit in the initial static snapshot ($Status_i^0, P_{gen_i}^0, P_{max_i}^0$, and $P_{min_i}^0$) are extracted. According to the availability of SCADA information for the unit, two approaches are then followed.

If the SCADA information is not available for the unit (for example due to the technical limitations of the SCADA system), the final status of the unit would be kept fixed ($Status_i^* = Status_i^0$). In this case, and if the unit status is OFF, the unit output power is set to zero ($P_{gen_i}^* = 0$) and the minimum and maximum practical powers of the unit are set equal to their initial values ($P_{min_i}^* = P_{min_i}^0$, $P_{max_i}^* = P_{max_i}^0$).

If the unit status is ON, its operating point is set equal to its initial value ($P_{gen_i}^* = P_{gen_i}^0$), and its maximum and minimum practical active powers are calculated based on (3.1) and (3.2), respectively.

$$P_{max_i}^* = \max(P_{max_i}^0, P_{gen_i}^* + \epsilon) \tag{3.1}$$

$$P_{min_i}^* = \max\{\min(P_{min_i}^0, P_{gen_i}^* - \epsilon), 0\} \tag{3.2}$$

These equations state that the modified maximum practical power ($P_{max_i}^*$) is equal to the maximum value of the initial maximum practical power ($P_{max_i}^0$) and unit operating point plus a small positive value ($P_{gen_i}^* + \epsilon$). A similar approach is followed for the minimum value. The maximizing function in (3.2), prevents the resulting value from becoming negative. These equations revise P_{max} and P_{min} of a generating unit based on P_{gen}. This is important for maintaining a stable and constant frequency at the beginning of the dynamic simulation and also matching the simulated and measured values of the generating units power changes.

However, if SCADA information is available, initially, $Status_i^{SCADA}$, $P_{gen_i}^{SCADA}$, $P_{max_i}^{SCADA}$, and $P_{min_i}^{SCADA}$ should be extracted based on (3.3)–(3.6), respectively.

$$Status_i^{SCADA} = \begin{cases} OFF; \ \overline{P_{gen_i}^{SCADA}(t)} < \alpha \\ ON; \ \overline{P_{gen_i}^{SCADA}(t)} \geq \alpha \end{cases} \tag{3.3}$$

$$P_{gen_i}^{SCADA} = \overline{P_{gen_i}^{SCADA}(t)}; \quad -T_a \leq t \leq 0 \tag{3.4}$$

$$P_{max_i}^{SCADA} = \max\left(P_{gen_i}^{SCADA}(t)\right); \forall t \tag{3.5}$$

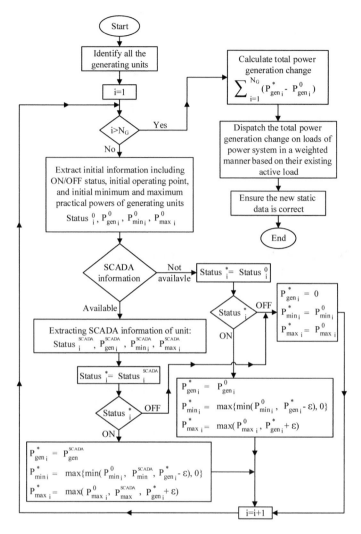

Fig. 3.2 Method for static data correction in the initial measures stage

$$P_{min_i}^{SCADA} = min\left(P_{gen_i}^{SCADA}(t)\right); \forall t \tag{3.6}$$

Equation (3.3) states that if the average power of a generating unit during the incident period $(\overline{P_{gen_i}^{SCADA}(t)})$ is smaller than a small positive value (α), the unit is considered to be OFF, otherwise, the unit is assumed to be ON. α should not be very small so that it can eliminate the possible error of the recorded data. In (3.4), $P_{gen_i}^{SCADA}$ is calculated by averaging its recorded generation power in T_a seconds time interval before the incident. The value of T_a should not be selected too large because the generating powers of the units vary continuously. In (3.5) and (3.6), the maximum and minimum power values recorded for the generating units active powers by the SCADA system, are considered as $P_{max_i}^{SCADA}$ and $P_{min_i}^{SCADA}$, respectively.

The ON and OFF status of a unit is determined based on SCADA (Status$_i^*$ = Status$_i^{SCADA}$), if available. In this case, and if the unit status is OFF, all formulations and explanations are the same as the situation when the SCADA information is not available. Besides, if under this circumstances, the status is ON, P_{gen} is set equal to the generation value based on SCADA data ($P_{gen_i}^* = P_{gen_i}^{SCADA}$) and P_{max} and P_{min} are calculated based on (3.7) and (3.8), respectively. The logic and explanations of these equations are similar to (3.1) and (3.2). The only difference is that $P_{max_i}^{SCADA}$ and $P_{min_i}^{SCADA}$ are added to the formulation to further improve the P_{max} and P_{min}.

$$P_{max_i}^* = \max(P_{max_i}^0, P_{max_i}^{SCADA}, P_{gen_i}^* + \epsilon) \tag{3.7}$$

$$P_{min_i}^* = \max\{\min(P_{min_i}^0, P_{min_i}^{SCADA}, P_{gen_i}^* - \epsilon), 0\} \tag{3.8}$$

The above correction process is repeated and carried out for all units ($i > i_G$). Following that, the modified static information should be balanced. For this purpose, initially, the total amount of power generation change due to modification of the generating units operating points is calculated based on (3.9). In order to balance the modified static information, this total power change is distributed among all power system loads in a weighted manner based on their existing active values.

$$\text{Total Power Change} = \sum_{i=1}^{N_G} \left(P_{gen_i}^* - P_{gen_i}^0 \right) \tag{3.9}$$

After balancing the total power system load and generation, it must be ensured that the modified static data is correct which means that the load flow based on the modified static data converges and the operating point of the unit selected as the slack generator is equal to the expected value. Otherwise, the loads active power changes should be corrected.

3.7 Quantitative Validation

For quantitative analysis on the matching performance of the validation process, four indices, namely (1) the average difference, (2) the absolute average difference, (3) the highest positive difference, and (4) the highest negative difference, are introduced in (3.10)–(3.13), respectively.

$$\text{Err}_{ave} = \frac{\sum_{t=t_i}^{t_f}(f_s(t) - f_r(t))}{\sum_{t_i=0}^{t_f} 1} \tag{3.10}$$

$$\text{Err}_{abs} = \frac{\sum_{t=t_i}^{t_f}|f_s(t) - f_r(t)|}{\sum_{t_i=0}^{t_f} 1} \tag{3.11}$$

$$\text{Err}_{max+} = \max_{t=t_i:t_f} \{\max(f_s(t) - f_r(t)), 0\} \tag{3.12}$$

$$Err_{max-} = min\{\min_{t=t_i:t_f}(f_s(t) - f_r(t)), 0\} \tag{3.13}$$

In which, t is time. The period under consideration is from t_i to t_f (from -10 to 120 s, zero being the incident time, in the investigated case study). f_r and f_s are the measured and the simulated signals of the frequency.

Err_{ave} and Err_{abs}, represent the average and absolute average of the difference between the simulated and measured values, respectively. Err_{ave} neutralizes the positive and negative values of the error and only shows the average matching that can be positive or negative (+ and − means, in general, larger and smaller simulated values as compared to the measured signal, respectively); whereas Err_{abs} expresses the absolute value of the error by adding all the error values, positively, which obviously always leads to a positive value. Err_{max+} and Err_{max-} represent the highest positive and negative differences between the simulated and the measured values, respectively. The maximum and minimum functions in (3.12) and (3.13) prevent Err_{max+} and Err_{max-} from becoming negative and positive, respectively.

As a general index for the selection of the best option in the internal loop of the framework, Err, which is the weighted average of the four previously introduced indices, is proposed as (3.14). Some thresholds as detailed in (3.15) have to be met.

$$Err = \left(\frac{|Err_{ave}|}{\alpha_{ave}} + \frac{Err_{abs}}{\alpha_{abs}} + \frac{Err_{max+}}{\alpha_{max+}} + \frac{|Err_{max-}|}{\alpha_{max-}}\right)/4 \tag{3.14}$$

$$|Err_{ave}| \leq \alpha_{ave}, Err_{abs} \leq \alpha_{abs}, Err_{max+} \leq \alpha_{max+}, Err_{max-} \geq -\alpha_{max-} \tag{3.15}$$

The choice of the values for the thresholds of each index is entirely case dependent. In the case study, as detailed in Chap. 4, 0.01, 0.02, 0.04 and 0.04 Hz are used for $\alpha_{ave}, \alpha_{abs}, \alpha_{max+}$ and α_{max-}, respectively. The option (regarding load parameters) with the least Err index which satisfies all of these thresholds is selected as the best option of simulations, in the internal loop of the framework,

In calculating the Err parameter, the positive value of each index is divided by its specified threshold which leads to a similar weighting of all the indices. After summing up these four values, the total is divided by 4 in order to have a positive and weighted value of Err. If each of the indices satisfies its threshold, the Err would be less than one. However, Err being less than one does not guarantee that all the indices are less than their specified thresholds.

This Err index is a weighted value of four different indices, which considers various aspects of matching; however, other conventional formulas such as $\sum(f_s(t) - f_r(t))^2$ can be used for assessing the quantitative matching of simulated and measured signals.

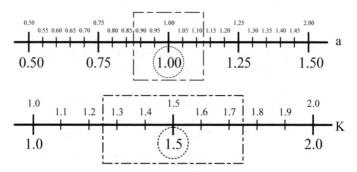

Fig. 3.3 Schematic of executing the simulations in the typical ranges of the voltage and frequency dependency of load (the hypothetical initial best option is "a = 1.0, K = 1.5")

3.8 Frequency and Voltage Dependency of Loads

Some details on the dependency of a load on voltage and frequency were already provided in Sect. 2.5.2. Two parameters, namely "a" and "K", were mentioned as the affecting parameters. These are not normally available and may change frequently due to the change in the composition of the power system load. That is why successive simulations are proposed to be performed for some typical ranges of these parameters. In order to reduce the number of such simulations for each incident, simulations are performed in two stages with large and small discrete steps of "a" and "K".

Therefore, in the first stage, the simulations are performed for "a" parameter from 0.5 to 1.5 with the step of 0.25 (5 values), and "K" parameter from 1 to 2 with the step of 0.5 (3 values). After choosing the best initial values among these 15 simulations, based on the matching indices (see Sect. 3.7) and considering the priority of reasonable load parameter values, in the second stage, the simulations will be performed for steps of 0.05 and 0.1 for "a" and "K" around the values of the best option, respectively (for each parameter, up to 5 values). Finally, the best option for each incident will be selected based on the matching indices among these up to 25 simulations. Figure 3.3 shows the schematic of this process.

Investigating the validity and effectiveness of the described generalized framework in this chapter on a standard test system is not possible due to the lack of having the measured behaviour of the system, for comparing with the simulated behaviour. Also, many steps of the framework, as explained in this chapter, are aimed at solving practical problems in an actual grid. Accordingly, in the next chapter, the Iranian power grid is selected as a case study in order to demonstrate the procedure of implementing the framework and to show its effectiveness for validating the frequency behaviour of a large and real power system.

Chapter 4
Implementation and Numerical Results

Abstract In this chapter, the proposed validation framework in Chap. 3 is applied to a large and interconnected *test grid* by modelling all known parameters affecting the frequency behaviour, including the parameters related to the active load and the generating units and also implementing the improving measures of the framework. This chapter aims to illustrate the step-by-step implementation of the validation framework for a large power system and demonstrates the effectiveness of the framework, its methods and formulations. It is concluded that for the investigated *test grid*, first approach (using existing models) is not sufficient to validate the frequency behaviour due to non-modelling of all affecting parameters. So after identifying the prominent non-modelled affecting parameters, a new generic model is selected and the validation process is successfully performed using the second approach (the new generic model). It should be mentioned that DSATools and its associated modules of PSAT and TSAT are used as the basic simulation tools.

Keywords Model validation · Implementation · DSATools · Frequency analysis
Frequency behaviour · System-Wide modelling · Governor parameters
Load parameters · Frequency stability · Frequency response

4.1 The Test Grid

The *test grid* with a total installed capacity of 76.4 GW and a total annual generated energy of 289 TWh is one of the world largest electricity grids (15th, in terms of installed capacity). The total installed capacity of different generation types in this grid is shown in Fig. 4.1. The total practical active power of the generating units at the grid peak load condition is 59.2 GW for a load of 52.2 GW. In this grid, the total capacity of transmission (400 and 230 kV) and the sub-transmission (63, 66, and 132 kV) power stations are 141,837 and 99,934 MVA, respectively. Moreover, the total length of the transmission and sub-transmission lines are 51,986 and 71,616 km, respectively.

© The Author(s), under exclusive license to Springer Nature Singapore Pte Ltd. 2019   37
H. Seifi and H. Delkhosh, *Model Validation for Power System Frequency Analysis*,
SpringerBriefs in Energy, https://doi.org/10.1007/978-981-13-2980-7_4

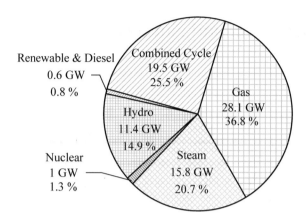

Fig. 4.1 Total installed capacity of different generation types in *test grid*

For the period of 1961–1979, 8 hydro units, each with an installed capacity of 65 MW (one with zero droop parameter) were responsible for frequency control; one in an automatic way (through governor) and the rest on manual control by the operator. The reference incident was then 150 MW, as the largest installed unit.

From 1979–2007, four new hydro units were added for this task, but still under manual control. The reference incident was increased to 440 MW, as the largest installed unit.

Some regulations were developed in 1997 which set the 49.7–50.3 Hz as the acceptable frequency range. In this range, some generating plants were specified to control the frequency, if a lasting (say more than 10 or 20 min) out-of-range frequency deviation happens. Even some load points were advocated for such a task so that by their interruptions, the frequency can be stabilized upon a major disturbance.

In that time, the automatic part of the task was still based on one single unit and the hydro units were the only generation type which did the frequency control job. On the other hand, the system was rapidly expanded and even connected to the neighbouring countries. That is why the tools were considered insufficient and the regulations were needed to be revised.

In 2007, due to a major restructuring of the *test grid*, the electric market made the trade on ancillary services, such as the one regarding the frequency control, possible. Now the gas turbine units, thermal and combined cycle ones are also able to be involved in frequency control, provided they are somehow paid. This has improved the frequency behaviour of the system to a large extent.

4.2 Initial Measures

Based on what was described in Chap. 3 (Sect. 3.5), some initial measures should be taken as follows:

4.2.1 Base Data Collection

Initially, some base data (different parameters value related to turbine-governor, generator, excitation system, etc.) should be collected in order to be used in the modelling procedure of the simulation. Some, mainly related to the governor, are as follows:

- Participation parameter for primary frequency control (PFP), either zero or 1, for all generating units. Transmission operator should provide such information. For the system under study, hydro units act as the secondary control. Therefore, this parameter is zero for such units. In the *test grid,* most of the thermal units are assumed to be active in primary control (PFP $= 1$). If for some specific reasons, a generating unit is inactive for such a task, its behaviour can be modelled by a high value of droop parameter in the simulation environment,
- Droop (R) and Dead Band (DB), as available. R of 5% and 4%, as well as DB of 50 MHz, are common for the *test grid*, as provided by the transmission operator,
- Maximum and minimum practical powers of the generating units. These have to be provided by the transmission operator, too. However, care should be taken to set, especially, the maximum values for the instant of the incident, as the maximum practical generation is dependent (especially for gas turbine units) on the environmental conditions. It should be noted that inaccurate values may be adjusted based on the methodology as described in Sect. 3.6,
- Activity Range (AR) of the units. The governors of some units are equipped with the capability of defining AR. If so (as provided by transmission operator), they should be properly modelled. For the *test grid*, some units are equipped so, and therefore are modelled,
- Frequency Ramp Rate (FRR) of the units have to be collected and modelled for all units. These may be different from the normal or fast ramp rates of the units. For the gas turbine units available in the grid under study, the normal and fast values are 11 and 30 MW/min, respectively,
- Inertia constant (H) of the generating units,
- Different time constants of the turbines and governors.

Table 4.1 Existing generator and turbine-governor models used for different generating unit types

Type	Number	Generator Model		Turbine-Governor Model			
		GENROUE	GENSAL	GAST	TGOV1	IEEEG1	IEEEG2
Hydro	50	0	50	0	0	0	50
Gas	114	114	0	114	0	0	0
Steam	58	58	0	3	26	29	0
CC#-Gas	88	88	0	86	0	2	0
CC-Steam	42	42	0	19	23	10	0
Total	**352**	**302**	**50**	**222**	**39**	**41**	**50**

Combined Cycle

4.2.2   Collecting Dynamic Information

Regarding the model and the parameters values of the generators, excitation systems, turbine-governors, and the power system stabilizers of the generating units, the data is provided by Iran Grid Management Company (IGMC). For the studies conducted in this monograph, the generator and turbine-governor models are essentially required. The relevant data for different generating unit types are shown in Table 4.1. Also, only the static models of transmission lines and transformers are used, which are sufficient for our purpose.

4.2.3   Selecting Incidents, Extracting the Closest Snapshot and Measured Signals

Regarding the three aforementioned initial measures, Table 4.2 presents the selected incident description and also the availability of static and SCADA data. As it can be seen, incidents are selected from various sizes and types of frequency decreasing (more frequent in the *test grid*) and increasing. The information such as a snapshot of static data close enough to the incident time, measured signals (frequency and active powers of nearly 250 generating units, participating in primary frequency control) by the SCADA system during the incident (from 2 min before the incident up to 10 min after that), the information about how the incidents have occurred, and also the subsequent emergency actions are available for all incidents.

4.2.4   Static Data Correction

The proposed static data correction method (see Sect. 3.6) is applied successfully on initial static data for each incident. Its overall results are presented in Table 4.3. The results show that despite the closeness of the selected snapshots to the incidents, the

Table 4.2 Description of the selected incident including static and SCADA data

Incident	Type	Size (MW)	Date	Time	Snapshot time	Load (GW)	SCADA data
1	Generation trip	750´	May 2016	11:49	11:35	30.5	Available
2	Generation trip	1050	February 2016	13:50	12:32	29.3	Available
3	Generation trip	280	May 2016	16:48	15:17	21.8	Available
4	Load trip	1600	June 2016	10:51	07:17	31.0	Available

Table 4.3 Results of implementing static data correction method–overall view

Incident	Positive change of P_{gen} (MW)	Negative change of P_{gen} (MW)	Change of P_{min} (MW)	Change of P_{max} (MW)	# of units from ON to OFF	# of units from OFF to ON
1	678	132	−412	+1190	0	1
2	992	522	−620	+1228	1	4
3	993	223	−438	+813	0	3
4	4384	317	−1039	+701	1	8

actual operating points of most units would differ from the initial values in snapshots. It is obvious that the correction of these parameters is very effective in active power change of the units during the incident and will result in better adaptation of the simulated and measured signals. The low number of units status change (from ON to OFF and vice versa) indicates the relatively high accuracy of the unit status in the initial static data, the main reason of which is the closeness of the snapshots to the incidents. The small increase in these values for the 4th incident, in which the snapshot is farther from the incident, is an indication of its correctness. Since the minimum practical generated power for most of the power plants is not a logical implemented limit and is just considered for the sake of conservatism by the operators, the lower value of the unit generation compared to this amount is not far from the mind. On the other hand, the maximum practical power of a generating unit is variable and often depends on the ambient temperature (especially for gas turbine units); therefore, using its average value (for example, the average monthly value) may create an error relative to the actual value. Obviously, corrections of P_{min} and P_{max} would help to match the simulated behaviour with the measured value for load and generation type incidents, respectively.

4.2.5 Incidents Details

As noted, all the information about the size of selected incidents and how they have occurred are available. Moreover, SCADA data is used so as to identify the emergency actions (mostly done by using hydro units of the *test grid*) after the incident. A time frame of 120 s (the time in which frequency usually is brought back to around nominal value) is considered for the validation process, although a different time frame may be considered depending on the system. Various types of emergency actions (either manual or automatic) may have happened which need to be thoroughly identified using the SCADA system. These have to be properly modelled in the simulation, based on the capabilities provided by the simulation tool.

Figure 4.2 shows different types (manual action, joint control system action, and manual action including small governor action at the beginning) of the identified active power changes of hydro units and their implementation as emergency actions in the simulations. Although implementing this measure needs high effort, it is vital for extending the time frame of the validation process. It should be noted that as in this measure, all of the hydro units power changes are handled as emergency actions in the simulation, the validation process is only concentrated on the turbine-governor of thermal units, which are the main participating units in primary frequency control for the *test grid*.

4.2.6   Correcting the Pre-incident Frequency

The small deviation of pre-incident frequency from its nominal value affects the frequency validation process significantly, mostly due to nonlinear effect of the dead band on droop curve. Accordingly, the pre-incident frequency of the power system at the simulation environment has to be adjusted equal to the measured frequency with the help of a moderate ramp (increase or decrease, whatever required) of power system load. The measured pre-incident frequencies for the four aforementioned incidents are 49.960, 50.137, 49.950, 50.025 Hz, respectively.

4.2.7   Determining the Typical and Acceptable Range of Active Load Parameters

Dynamic simulations of each incident are performed for the typical range of frequency and voltage dependency of active loads. In order to reduce the number of dynamic simulations, they are performed in two stages with large and small steps of the affecting load parameters. Thus, in the first stage (large steps), the simulations are performed for the "a" parameter from 0.5 to 1.5 by steps of 0.25, and the "K" parameter from 1 to 2 with steps of 0.5.

After choosing the best initial values based on the matching indices among these simulations, in the second stage (small steps), the simulations are performed for the steps of 0.05 and 0.1 for "a" and "K" around the values of the best option, respectively. Some details were provided in Sect. 2.5.2 and 3.8.

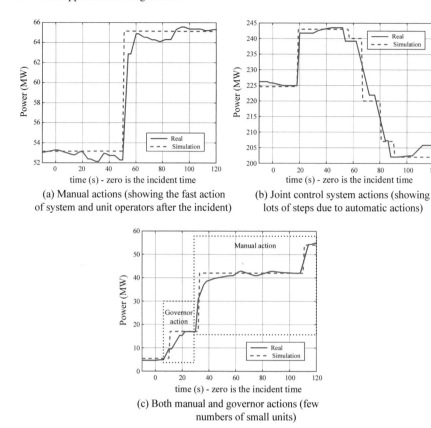

(a) Manual actions (showing the fast action of system and unit operators after the incident)

(b) Joint control system actions (showing lots of steps due to automatic actions)

(c) Both manual and governor actions (few numbers of small units)

Fig. 4.2 Identified and implemented emergency actions via SCADA data

4.3 First Approach: Existing Models

Initially and as a first approach, the maximum capabilities of existing turbine-governor models are tried. In this section, this approach is initially applied to the *test grid*. In case of failure and unacceptable validation performance, the second approach as detailed in Sect. 4.4, would be tried.

4.3.1 Existing Models

Existing standard turbine-governor models used for the generating units of the *test grid* are presented in Fig. 4.3.

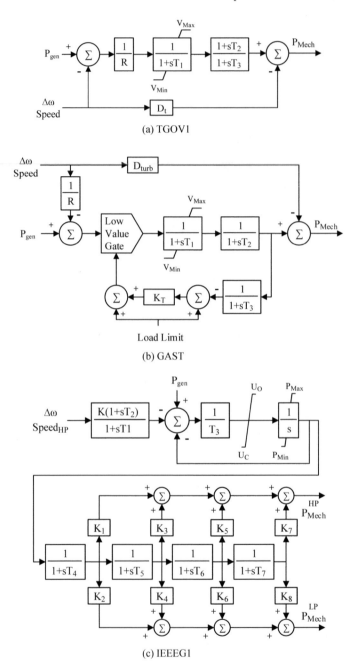

Fig. 4.3 Existing turbine-governor models

Table 4.4 Capabilities of existing turbine-governor models

Model	% of participation	Droop	DB	P_{max}	P_{min}	Activity Range	Frequency Ramp Rate
TGOV1	8%	R	×	V_{Max}	V_{Min}	×	×
GAST	86%	R	×	V_{Max}	V_{Min}	×	×
IEEEG1	6%	1/K	×	P_{Max}	P_{Min}	×	U_C, U_O
IEEEG2	0%	Hydro units–Simulated for emergency actions					

In TGOV1 (see Fig. 4.3a), which is used for steam units of the *test grid*, typical values for T_1, T_3, T_2/T_3, D_t, and R are 0.5 s, 5–9 s, 0.3, 0, and 0.05, respectively.

In GAST (see Fig. 4.3b), which is used for gas turbine and combined cycle units of the *test grid*, typical values for T_1, T_2, T_3, K_t, D_t, and R are 0.4 s, 0.1 s, 3 s, 2, 0 and 0.05, respectively.

IEEEG1 (see Fig. 4.3c) is a model developed by IEEE and after tuning its parameters, can be applied to various types of turbines. Typical values for T_1, T_2, T_3, T_4, T_5, T_6, T_7, K_1, K_2, K_3, K_4, K_5, K_6, K_7, K_8, K, U_O, and U_C are 0–0.5 s, 0–10 s, 0.04–1 s, 0–1 s, 0–10 s, 0–10 s, 0–10 s, 0–1, 0, 0–0.5, 0–0.5, 0–0.35, 0–0.55, 0–0.3, 0–0.3, 20, 0.01–0.3 p.u./s, and 0.01–0.3 p.u./s, respectively. It should be mentioned that the sum of K_1, K_3, K_5, and K_7 as well as the sum of K_2, K_4, K_6, and K_8, should be one. For hydro turbines, IEEEG2 governor is used. Capabilities of these turbine-governor models for modelling different affecting parameters on frequency behaviour are demonstrated in Table 4.4.

The following results can be concluded about the existing models:

- All the models have properly modelled the time constants involved and different mechanical characteristics of different generating unit types,
- All the models have considered droop (R) curve without Dead Band (DB),
- All the models have parameters for considering the maximum and minimum practical active powers (P_{max}, P_{min}),
- None of the models is capable of modelling the activity range (AR) parameter,
- The maximum frequency ramp rate (approximately 0.15 p.u./min or 0.012 p.u./s for many of the *test grid* gas turbine generating units) can only be considered in IEEEG1 model, through the physical maximum opening and closing speeds of the steam valve (approximately 0.3 p.u./s).

It is worth mentioning that inertia (H) is considered in the generator model.

4.3.2 First Incident Simulation Results

In this section, the frequency behaviour of the *test grid* for the first incident is investigated using the existing turbine-governor models, based on the modified static

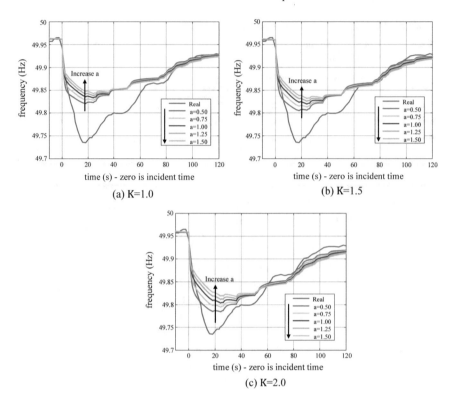

Fig. 4.4 Simulated frequency of the first incident with large steps of load voltage and frequency dependency parameters using approach #1

data, with the correction of the pre-incident frequency and the implementation of emergency actions. The simulation results are shown in Fig. 4.4 for the large load parameters steps (5 "a" and 3 "K" values). A qualitative comparison of these results with the measured frequency of the grid shows that none of the simulations has a fairly consistent adaptation to the measured frequency and it seems that some of the affecting parameters, especially for the first 60 s after the incident, have not been properly identified and modelled. The best option has the lowest value of "a" and "K", which still does not match to the measured frequency profile.

In order to investigate the simulations quantitatively, the matching indices for these 15 simulations are calculated and the results with the specified thresholds are depicted in Fig. 4.5. As can be seen, none of these 15 cases has met all of the minimum requirements, and among them, the best option, with the smallest Err index and satisfying the highest number of minimum conditions, is for "a = 0.5 and K = 1".

Therefore, as the first approach has generally failed to provide a good basis for the validation process of the first incident, other incidents would not be checked.

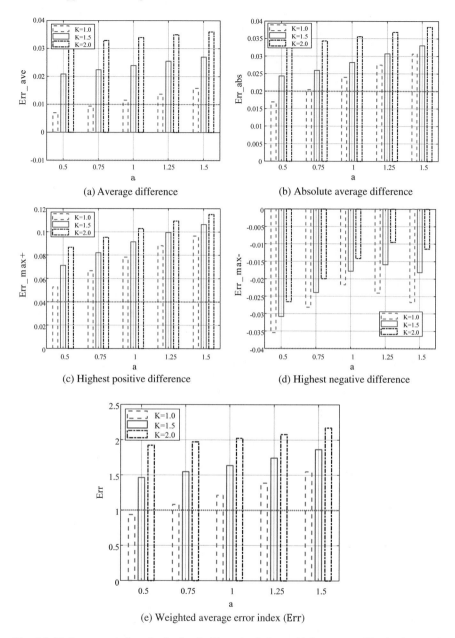

Fig. 4.5 Various error indices for the first incident simulations with large steps of load voltage and frequency dependency parameters using approach #1

Moreover, as even the performance of the best option in large steps of load parameters is not acceptable, the simulations for the small steps are not performed.

Some details will be given below, to identify new affecting parameters using the comparison of the simulated and the measured active powers of the units, in order to develop a basis to choose a new and proper generic model of turbine governors for the validation process (as detailed in Sect. 4.4).

4.3.3 Identification of New Affecting Parameters

The measured and simulated active powers of some generating units are compared in Fig. 4.6, in order to identify new affecting parameters. According to Table 4.4, three non-modelled affecting parameters in the existing models are Dead Band (DB), Activity Range (AR), and Frequency Ramp Rate (FRR).

All the existing models have considered the droop characteristics without the dead band parameter. Although the frequency deviations of the incidents for the *test grid* is not very small (sometimes more than 0.2 Hz), the dead band parameter (usually 0.05 Hz in *test grid*) cannot be overlooked. Also, lack of dead band may lead to a change of unit generation (P_{gen}) due to the small deviation of pre-incident frequency from its measured value. This can also result in improper reaching or non-reaching to P_{max} and P_{min} of the unit during incidents, which in turn, would considerably affect the frequency behaviour of the whole system. These are evident in Figs. 4.6a and 4.6b (with the change of only the operating point) and Fig. 4.6c (change of the operating point and improper reaching to the maximum power).

Regarding the activity range, it should be noted that this is not a restricting constraint for the most of the generating units of the *test grid* and if necessary, can be taken into account indirectly through P_{max} and P_{min}, so that if $P_{gen} + AR$ is less than P_{max}, this value will be used instead of P_{max}, and if $P_{gen} - AR$ is greater than P_{min}, this value substitutes P_{min}.

Regarding the frequency ramp rate, almost 94% of the participating units in the primary frequency control (with TGOV1 and GAST models) do not model this parameter. Given that for the *test grid*, this logical limitation exists for many units (especially gas turbine units, most of which use the GAST model), this parameter can be one of the main reasons of the mismatch between the simulation result and the measured behaviour. It is worth noting that this parameter theoretically affects the simulation in the initial moments after the incident (say the first 60 s), which is compatible with our case (see Sect. 4.3.2). The effect of this parameter on the active powers of the generating units is evident in Figs. 4.6b and 4.6c.

Therefore, a new model should be selected in such a way that it provides the facility to model dead band and frequency ramp rate directly and allows modelling of the activity range parameter, at least, indirectly (say through P_{max} and P_{min}).

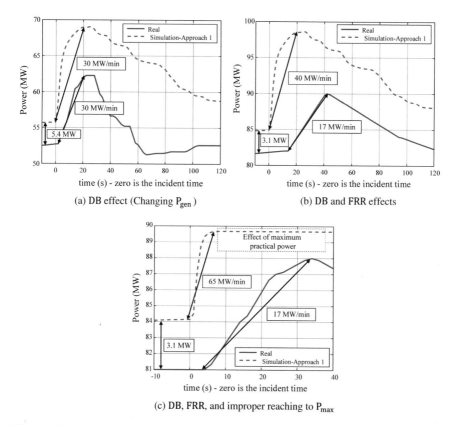

(a) DB effect (Changing P_{gen})

(b) DB and FRR effects

(c) DB, FRR, and improper reaching to P_{max}

Fig. 4.6 Comparing simulated and measured active powers of some generating units (first incident–approach #1

4.4   Second Approach: New Generic Model

As shown, the existing models were inadequate to validate the frequency behaviour of the *test grid* and hence the use of the second approach is inevitable. In this approach, a new model with the capability of modelling all the identified affecting parameters is utilized for model validation.

4.4.1   New Generic Model

As aforementioned, non-observation of the frequency ramp rate and the dead band of thermal units were identified as the main factors of the validation process failure,

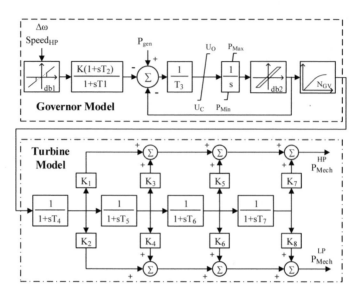

Fig. 4.7 New selected standard turbine-governor model-WSIEG1

using the first approach. Therefore, in the second approach, WSIEG1 standard model (the extended version of the IEEEG1 standard model with the capability of modelling these parameters) is selected as the new generic model. As shown in Fig. 4.7, the turbine model of WSIEG1 has no difference with that of the IEEEG1, and only a few blocks have been added to the governor section. This governor can model droop (R), Dead Band (DB), maximum and minimum practical powers and Frequency Ramp Rate (FRR), through 1/K, db1, (P_{Max}, P_{Min}), and (U_O, U_C), respectively. Also, if necessary, the Activity Range (AR) can be integrated into P_{max} and P_{min} as described earlier. Therefore, if other parameters of existing models (such as time constants) can be somehow converted to the parameters of this model, it would be perfectly suitable for the studies of concern.

4.4.2 Converting Model Parameters

The parameters conversion formulations of the three existing models for the *test grid* to the new model parameters are shown in Table 4.5. It should be noted that in these equations, the parameters at the left and the right of the equality sign are, respectively, related to WSIEG1 model and the existing models. As it can be seen, the equations are fairly simple and often relate the time constants of two models to each other. In order to ensure the correctness of the model conversion process, the simulation of the second approach is performed without applying the values of

Table 4.5 Formulations for the conversion process of the models

TGOV1	GAST	IEEEG1
$K = 1/R$	$K = 1/R$	All the parameters are one by one the same
$T_3 = T_1$	$T_3 = T_1$	
$K_1 = T_2/T_3$	$T_4 = T_2$	
$K_3 = 1 - T_2/T_3$	$K_1 = 1$	
$T_5 = T_3$		

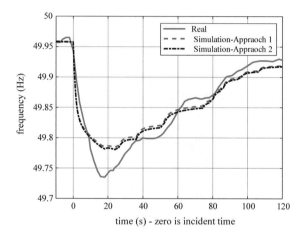

Fig. 4.8 Comparing simulation results of the first and the second approaches in order to evaluate the correctness of parameters conversion process

new parameters (DB, AR, and FRR). The results are compared with the simulation based on the first approach (best option) as shown in Fig. 4.8. They prove the proper conversion process.

4.4.3 First Incident Simulation Results

The simulations results for different load parameters, 5 values of "a" (the step of 0.25 from 0.5 to 1.5) and 3 values for "K" (the step of 0.5 from 1 to 2) are shown in Fig. 4.9. The proposed matching indices for these 15 simulations are calculated and the results are depicted in Fig. 4.10.

The qualitative investigation shows that the performed simulations based on this approach (new generic model) have relatively better matching with the measured frequency, compared to the first approach. By adding newly identified affecting parameters (DB, AR, and FRR), good matching is achieved, especially for the first 60 s. Considering the quantitative indices, three of these 15 cases ("a = 1, K = 1.5";

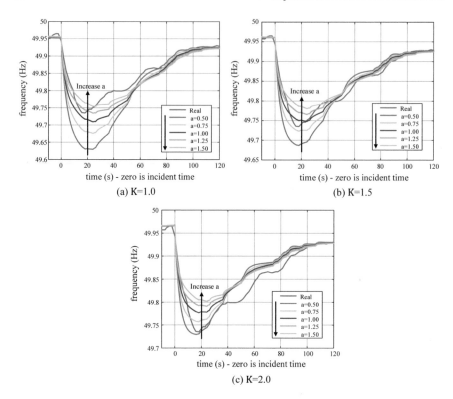

Fig. 4.9 Simulated frequency for the first incident with large steps of load voltage and frequency dependency parameters using approach #2

"a = 1.25, K = 1.5"; and "a = 1.5, K = 1") have satisfied all the minimum thresholds, among which the best option, with the lowest value of Err, is "a = 1, K = 1.5".

The next step is to perform the simulation with the small steps of load parameters around this best option. Accordingly, a total of 25 simulations with 5 "a" values (with a step of 0.05 around 1, which means 0.9, 0.95, 1, 1.05, and 1.1) and 5 "K" values (with a step of 0.1 around 1.5, which means 1.3, 1.4, 1.5, 1.6, and 1.7) are performed, the results of which are presented in Fig. 4.11. Also, the calculated matching indices are demonstrated in Fig. 4.12. The option with the lowest Err is for "a = 1, K = 1.5", same as before. After this stage, other incidents should also be investigated in order to ensure that the approach is completely successful.

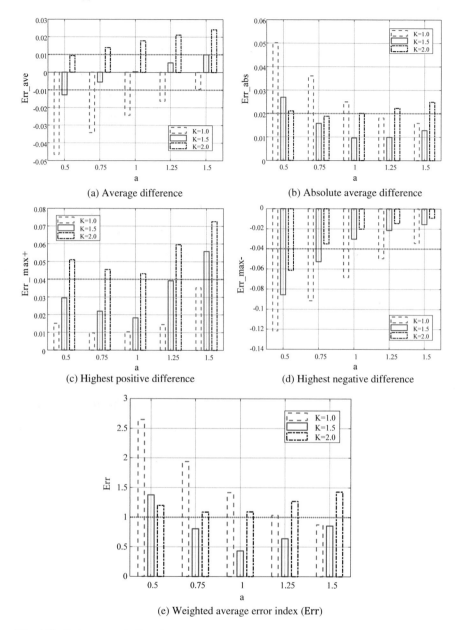

Fig. 4.10 Various error indices for the first incident simulations with large steps of load voltage and frequency dependency parameters using approach #2

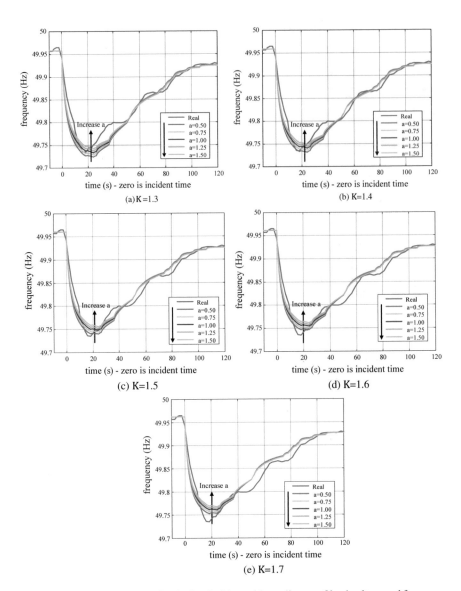

Fig. 4.11 Simulated frequency for the first incident with small steps of load voltage and frequency dependency parameters using approach #2

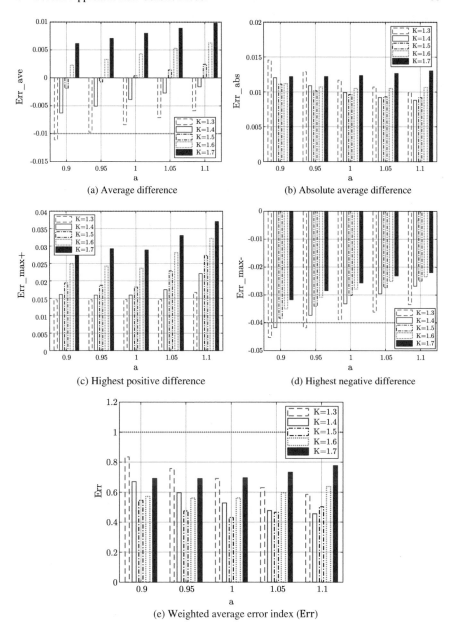

Fig. 4.12 Various error indices for the first incident simulations with small steps of load voltage and frequency dependency parameters using approach #2

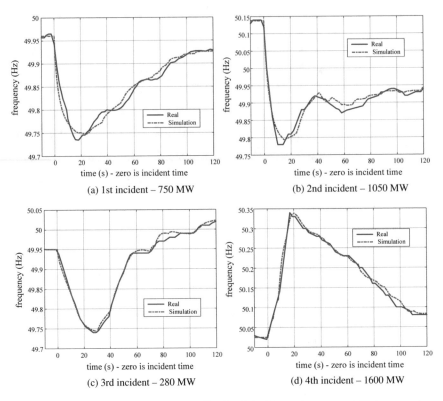

Fig. 4.13 The best simulation frequency profile of each incident, using approach #2, in comparison with the measured frequency

4.4.4 Other Incidents Simulation Results

With the steps as discussed above, the second approach results in an acceptable validation process for the first incident. In this section, this process has been tried for other incidents with successful and satisfactory results.

The best option frequency profile for each of these incidents is presented in Fig. 4.13. The quantitative indices for these best options are also depicted in Fig. 4.14. For all incidents, the simulated frequency profile is quantitatively and qualitatively matched to the measured frequency with reasonable load parameters. Therefore, the model validation is finished successfully using the second approach.

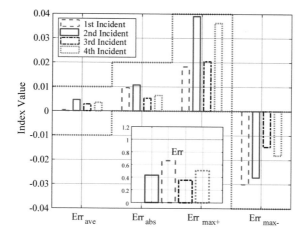

Fig. 4.14 Various error indices for each incident best option

The effect of different affecting parameters (load and generating units parameters) and the improving measures of the framework on the validation process (validated frequency profile of the first incident) are investigated in the next chapter through developing different scenarios.

Chapter 5
Detailed Sensitivity Analysis

Abstract In this chapter, the effects of different affecting parameters and improving measures on frequency validation are investigated by developing various scenarios based on the validated simulation. These scenarios are developed in three main groups which are the parameters of the loads (voltage and frequency dependency), the parameters of the turbine-governors (participation in primary frequency control, droop, dead band, practical powers, activity range, frequency ramp rate, and various turbine and governor time constants), and the improving measures of the framework (pre-incident frequency correction, secondary and tertiary emergency actions, static data correction, and the use of accurate governor parameters). Qualitative and quantitative comparisons of these simulations with the measured frequency profile and also the validated simulation show that considering all of these parameters and improving measures are vital for successful power system model validation from the power system frequency point of view.

Keywords Model validation · Sensitivity analysis · DSATools
Frequency analysis · Frequency behaviour · Governor parameters effect
Load parameters effect · Validation measures effect · Frequency response
Frequency stability

5.1 Introduction

In Chap. 3, the framework for the validation process was provided. In Chap. 4, its implementation for a practical large-scale system was discussed. Specifically, it was shown that how a validated model might be achieved using a generic governor model with tuned parameters. In achieving such a goal, some initial measures were taken. Moreover, by assuming similar load characteristics for all nodes, the best relevant parameters ("a" and "K") were determined.

In this chapter, the numerical studies are extended to some scenarios in order to analyse the sensitivity of the validation performance with respect to various parameters and measures, as proposed and used in Chaps. 3 and 4. Such studies would provide some valuable information on enhancing the primary control performance of a system if decided to be done. Base scenario is considered as what was developed in Chap. 4 for incident 1, with approach 2 (see Sect.4.4), in which the load parameters were "$a = 1$ and $K = 1.5$".

5.2 Basic Parameters and Measures

The basic parameters and measures are evaluated as follows:

(a) Parameters:

- Voltage dependency of load ("a"),
- Frequency dependency of load ("K"),
- Primary Frequency Participation (PFP),
- Droop parameter (R),
- Dead Band (DB),
- Maximum and minimum practical powers (P_{max} and P_{min}),
- Generating unit Activity Range (AR),
- Frequency Ramp Rate (FRR),
- Various turbine and governor time constants (T).

(b) Measures:

- Pre-incident frequency correction,
- Secondary and tertiary emergency actions implementation,
- Static data correction implementation,
- The use of accurate governor parameters.

5.3 Scenarios

13 scenarios are proposed as detailed below. All are developed on the basis of the base scenario as already described:

- *Scenario 1*: Consider the voltage dependency of loads to be zero ($a = 0$),
- *Scenario 2*: Consider the frequency dependency of loads to be zero ($K = 0$),
- *Scenario 3*: In this scenario, it is assumed that all generating units are participating in primary frequency control (PFP $= 1$). For those units with already zero participation, the governor and turbine parameters are set as those for the already participating units,
- *Scenario 4*: Consider droop parameter (R) to be zero for all generating units. This scenario has the same effect as none of the generating units participating in frequency control (PFP $= 0$ for all of the units),
- *Scenario 5*: Consider Dead Band (DB) to be zero for all generating units,

- *Scenario 6*: Relax the maximum and minimum practical powers of all units (i.e. set $P_{max} = 1.0$ and $P_{min} = 0.0$),
- *Scenario 7*: Relax the Activity Range (AR) of all units by setting the respective values to P_{max} and P_{min},
- *Scenario 8*: Relax Frequency Ramp Rate (FRR) of all generating units by its setting to 1.0 p.u./s,
- *Scenario 9*: Set all of the time constants, as in WSIEG1 model (T_1–T_7), to zero,
- *Scenario 10*: Ignore the pre-incident frequency correction (see Sects. 3.5 and 4.2.6),
- *Scenario 11*: Ignore the secondary and tertiary emergency actions (see Sects. 3.5 and 4.2.5),
- *Scenario 12*: Ignore the static data correction method (see Sects. 3.5 and 4.2.4),
- *Scenario 13*: Consider R (for all units), DB (for all units), FRR (for steam units), FRR (for gas turbine units), H (for steam units) and H (for gas turbine units) to be 0.05, 0.05 Hz, 11 MW/min, 5 MW/min, 2.5 and 5.0 s, respectively, as typical values; instead of their actual values. Moreover, relax the Activity Range (AR) of all generating units.

5.4 Simulation Results

Figures 5.1, 5.2, and 5.3 depict the simulation results for the above scenarios along with the results of the base scenario. The matching indices formulated as (3.10)–(3.14) are also calculated and shown in Figs. 5.4 and 5.5. The indices, shown in Fig. 5.4, are for the case in which the results are compared with those of the measured behaviour, whereas the indices in Fig. 5.5 are for the case in which the results are compared with those of the best case for approach 2 (i.e., the base scenario).

The followings may be concluded about the results:

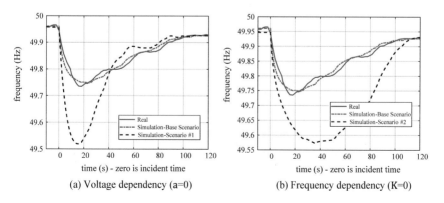

(a) Voltage dependency (a=0)   (b) Frequency dependency (K=0)

Fig. 5.1 Effect of ignoring the load parameters on the validated frequency profile

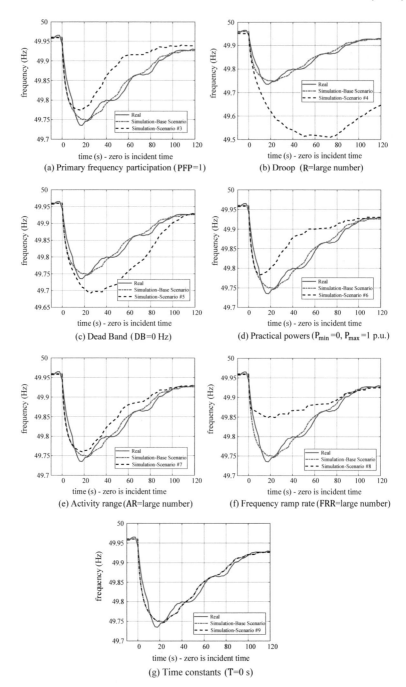

Fig. 5.2 Effect of ignoring the generating units parameters on the validated frequency profile

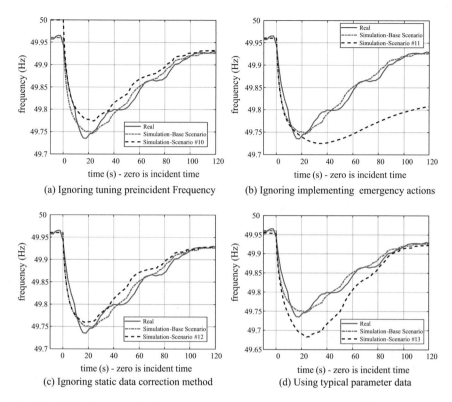

Fig. 5.3 Effect of Ignoring the improving measures of the framework on the validated frequency profile

- *Scenarios 1 and 2* (Figs. 5.1a and 5.1b): The results for these two scenarios show the great importance of observing the load characteristics, in any validation process,
- *Scenario 3* (Fig. 5.2a): The results of this scenario show some noticeable effects on the performance. However, the quantitative indices show that the increase of generating units participation from 44 to 100%, does not significantly affect the validation process. The main reason is that some other limitations (such as P_{gen}, P_{max}, P_{min}, FRR, etc.) may have prevented from achieving a noticeable better frequency performance,
- *Scenario 4* (Fig. 5.2b): The results for this scenario show how important the observing of the droop parameter is. In fact, by its ignorance, the frequency is merely controlled by the inertias, voltage and frequency dependency of loads and the secondary and the tertiary emergency actions,
- *Scenario 5* (Fig. 5.2c): As the results show for this scenario, the importance of observing the dead band is evident, especially after the 20th second onward,
- *Scenario 6* (Fig. 5.2d): If P_{max} and P_{min} are relaxed, the results for this scenario show some quite different results. Therefore, they cannot be ignored,

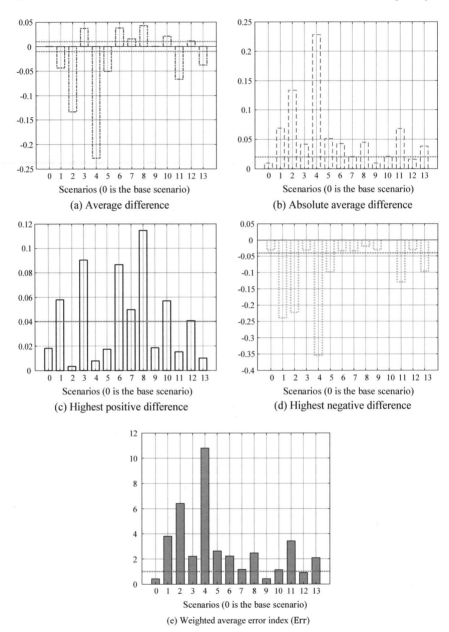

Fig. 5.4 Various error indices of different scenarios as compared to the measured frequency profile

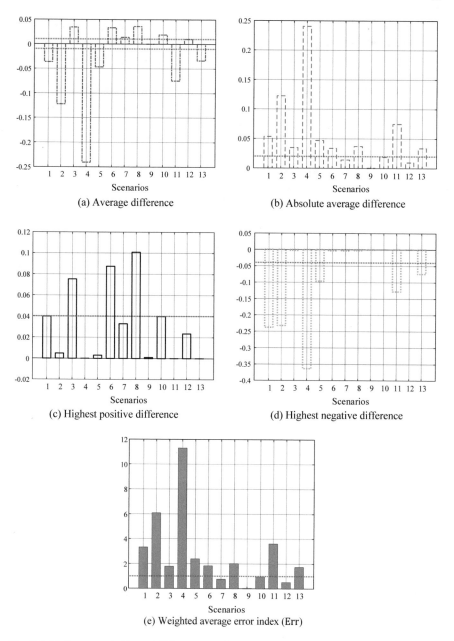

Fig. 5.5 Various error indices of different scenarios as compared to the best simulation

- *Scenario 7* (Fig. 5.2e): If the AR s are ignored in the modelling stage, somewhat different, but not significant, results may be achieved. The main reason is the fact that in the system under study, there are not so many such generating units with constrained ARs,
- *Scenario 8* (Fig. 5.2f): As shown for this scenario, FRR has some dominant effects on the validation performance, especially from the 10th second to the 70th second and cannot be overlooked at all,
- *Scenario 9* (Fig. 5.2g): Due to the small values of the time constants, ignoring them does not significantly affect the validation process,
- *Scenario 10* (Fig. 5.3a): The results for this scenario show the importance of the pre-incident frequency correction; although this correction is insignificant. The main reason is due to the DB which makes the droop characteristics nonlinear,
- *Scenario 11* (Fig. 5.3b): Observing the emergency actions significantly affects the validation performance after the 20th second onward. If ignored, the process can be validated only for a shorter time frame,
- *Scenario 12* (Fig. 5.3c): Static data correction seems to have less noticeable effect in comparison with some other parameters. The reason is that some type of corrections may have been compensated with some others so that the overall effect is not significant,
- *Scenario 13* (Fig. 5.3d): The importance of these parameters is evident from the results for this scenario, so that good validation process may not be achieved by using some typical values for such parameters.

References

Some basic materials on various issues of power system operation and dynamics are covered in [27, 31, 43, 63], in which load frequency control is also discussed. Historical review of frequency issue is discussed in [2, 46, 48, 55]. Various types of operating reserves are mentioned in [13, 58].

Frequency control policies in UCTE, ENTSO-E, and some European countries (namely, Germany, Ireland and England) are described in [60, 61], [16–20] and [7, 12, 15, 62], respectively. EPRI [4, 21, 50–53] provide such information on NERC, whose objective is to ensure the reliability of the power grid in North America. The reports on the pricing of ancillary services by EPRI are given in [22–26, 28–30], while [32–36] give the reports on frequency control by Laurence Berkley Lab.

Classification of the validation process is outlined in [1]. Some component-based and hybrid validation algorithms and implementations are reviewed in [5, 6, 47, 54] and [3, 37, 44, 64, 65], respectively. Some experiences on system-wide model validation from the frequency perspective in real power grids are provided in [8, 40–42, 56, 57, 59]. Relevant data and information for the test grid are from [14, 39, 45]. Turbine-governor models are described in [38, 49]. Finally, some of the manuals of the used simulation tool (DSATools™) are [9–11].

[1] Allen E, Kosterev D, Pourbeik P (2010) July. Validation of power system models. In: Power and energy society general meeting, 2010 IEEE, pp 1–7. IEEE
[2] Blalock TJ (2003) The frequency changer era-interconnecting systems of varying cycles. IEEE Power Energy Mag 99(5):72–79
[3] Bo BO, Jin MA, Renmu HE (2009) A hybrid dynamic simulation validation strategy based on time-varied impedance. Autom Electr Power Syst 6, 003
[4] CERTS (2002) Frequency control concerns in the North American electric power system, Consortium for Electric Reliability Technology Solutions (CERTS).
[5] Chang G, Hatziadoniu C, Xu W, Ribeiro P, Burch R, Grady WM, Halpin M, Liu Y, Ranade S, Ruthman D, Watson N (2004). Modeling devices with nonlinear voltage-current characteristics for harmonic studies. IEEE Trans Power Deliv 19(4):1802–1811.
[6] Chen C, Zhou Z, Bollas GM (2017) Dynamic modeling, simulation and optimization of a subcritical steam power plant. Part I: Plant model and regulatory control. Energy Convers Manag 145:324–334

© The Author(s), under exclusive license to Springer Nature Singapore Pte Ltd. 2019
H. Seifi and H. Delkhosh, *Model Validation for Power System Frequency Analysis*,
SpringerBriefs in Energy, https://doi.org/10.1007/978-981-13-2980-7

[7] Consentec (2014) Description of load-frequency control concept and market for control reserves, Study commissioned by the German TSOs, ordered by 50 Hertz Transmission GmbH

[8] Decker IC, e Silva AS, da Silva RJG, Agostini MN, Martins N, Prioste FB (2010). System wide model validation of the Brazilian interconnected power system. In: Power and energy society general meeting, 2010 IEEE, pp 1–8. IEEE

[9] DSATools™, PSAT (Powerflow & Short circuit Assessment Tool), User Manual

[10] DSATools™, TSAT (Transient Security Assessment Tool), Model Manual

[11] DSATools™, TSAT (Transient Security Assessment Tool), User Manual

[12] EirGrid, 2015, EirGrid Grid Code, Version 6, Ireland

[13] Ela E, Milligan M, Kirby B (2011) Operating reserves and variable generation. Contract, vol 303, pp 275–3000

[14] Electric Power Industry Statistics, http://amar.tavanir.org.ir/en/, accessed: 19 Aug 2018

[15] Energy UK (2017) Ancillary services report 2017, UK

[16] ENTSO-E (2011) Deterministic frequency deviations—root causes and proposals for potential solutions

[17] ENTSO-E (2013) Deterministic frequency deviations-Initial findings report

[18] ENTSO-E (2013) Network code for requirements for grid connection applicable to all generators

[19] ENTSO-E (2013) Network code on load-frequency control and reserves

[20] ENTSO-E (2013) Supporting document for the network code on load-frequency control and reserves

[21] EPRI (1992) Impacts of governor response changes on the security of north american interconnections

[22] EPRI (1997) Cost of providing ancillary services from power plants, vol 1: a Primer

[23] EPRI (1997) Cost of providing ancillary services from power plants, vol 2: regulation and frequency response

[24] EPRI (1997) Cost of providing ancillary services from power plants, vol 3: operating reserve

[25] EPRI (1997) Fixed costs of providing ancillary services from power plants: reactive supply and voltage control, regulation, frequency response and operating reserve

[26] EPRI (1998) Mechanisms for evaluating the role of hydroelectric generation in ancillary service markets

[27] EPRI (1998) Dynamics of interconnected power systems tutorial: second edition, EPRI Technical Results

[28] EPRI (2000) Methodology for costing ancillary services from hydro resources

[29] EPRI (2001) Accommodating wear and tear effects on hydroelectric facilities operating to provide ancillary services

[30] EPRI (2002) Revenues from ancillary services and the value of operational flexibility

[31] Eremia M, Shahidehpour M (eds) (2013) Handbook of electrical power system dynamics: modeling, stability, and control (vol 92). Wiley, USA

[32] Ernest Orlando Lawrence Berkeley National Laboratory (2010) Dynamic simulation studies of the frequency response of the three U.S. interconnections with increased wind generation, Ordered by FERC

[33] Ernest Orlando Lawrence Berkeley National Laboratory (2010) Frequency control performance measurement and requirements, Ordered by FERC

[34] Ernest Orlando Lawrence Berkeley National Laboratory (2010) Power and frequency control as it relates to wind-powered generation, Ordered by FERC

[35] Ernest Orlando Lawrence Berkeley National Laboratory (2010) Review of the recent frequency performance of the eastern, Western and ERCOT Interconnections, Ordered by FERC

[36] Ernest Orlando Lawrence Berkeley National Laboratory (2010) Use of frequency response metrics to assess the planning and operating requirements for reliable integration of variable renewable generation, Ordered by FERC

[37] Hajnoroozi AA, Aminifar F, Ayoubzadeh H (2015) Generating unit model validation and calibration through synchrophasor measurements. IEEE Trans Smart Grid 6(1):441–449

[38] IEEE Task Force on Turbine-Governor Modeling (2013) Dynamic models for turbine-governors in power system studies, Sponsored by Power System Stability Subcommittee

[39] Iran Grid Management Co., Independent System Operator (ISO) of Iran, http://www.igmc.ir/en. Accessed: 19 Aug 2018

[40] Jun X, Dysko A (2013) UK transmission system modelling and validation for dynamic studies. In: Conference on IEEE/PES, ISGT EUROPE, Copenhagen, Denmark, pp 6–9

[41] Kou G, Hadley S, Liu Y (2014) Dynamic model validation with governor deadband on the eastern interconnection. Oak Ridge National Laboratory, The Power and Energy Systems Group, Oak Ridge, TN, USA, Tech. Rep. ORNL/TM-2014/40

[42] Kou G, Markham P, Hadley S, King T, Liu Y (2016) Impact of governor deadband on frequency response of the US Eastern Interconnection. IEEE Trans Smart Grid 7(3):1368–1377

[43] Kundur P, Balu NJ, Lauby MG (1994) Power system stability and control (vol 7). McGraw-hill, New York

[44] Ma J, Han DONG, Sheng WJ, He RM, Yue CY, Zhang J (2008) Wide area measurements-based model validation and its application. IET Gener Transm Distrib 2 (6):906–916

[45] Mashhadi MR, DB, MHJ, Ghazizadeh MS (2011) The impacts of capabilities and constraints of generating units on simultaneous scheduling of energy and primary reserve. Electr Eng 93 (3):117–126

[46] Mixon P (1999) Technical origins of 60 Hz as the standard AC frequency in North America. IEEE Power Eng Rev 19(3):35–37

[47] Murat D, Kosalay I, Gezer D, Sahin C (2015). Validation of hydroelectric power plant model for speed governor development studies. In: 2015 international conference on renewable energy research and applications (ICRERA), pp 278–282. IEEE.

[48] Neidhöfer G (2011) 50-Hz drequency [History] IEEE Power Energy Mag 4(9):66–81

[49] NEPLAN, Standard Dynamic Turbine-Governor Systems in NEPLAN power system analysis tool, turbine-governor models, Available Online www.neplan.ch

[50] NERC (2012) Frequency response initiative report-the reliability role of frequency response

[51] NERC (2012) Frequency response standard background document

[52] NERC (2015) Primary frequency, reliability guideline

[53] NERC (2016) Reliability Standards for the bulk electric systems of North America

[54] Olivenza-León D, Medina A, Hernández AC (2015) Thermodynamic modeling of a hybrid solar gas-turbine power plant. Energy Convers Manag 93:435–447

[55] Owen EL (1997) The origins of 60-Hz as a power frequency. IEEE Ind Appl Mag 3(6):8–14

[56] Pereira L, Kosterev D, Davies D, Patterson S (2004) New thermal governor model selection and validation in the WECC. IEEE Trans Power Systems 19(1):517–523

[57] Pereira L, Undrill J, Kosterev D, Davies D, Patterson S (2003) A new thermal governor modeling approach in the WECC. IEEE Trans Power Systems 18(2):819–829

[58] Rebours Y, Kirschen D (2005) What is spinning reserve. The University of Manchester, vol 174, p 175

[59] Stavrinos S, Petoussis AG, Theophanous AL, Pillutla S, Prabhakara FS (2010) Development of a validated dynamic model of Cyprus Transmission system. Power generation, transmission, distribution and energy conversion (MedPower 2010), IET.

[60] UCTE (2009) UCTE Operation Handbook_ Policy 1_ Load-Frequency Control and Performance

[61] UCTE (2009) UCTE_Operation Handbook_Policy 1_Appendix

[62] VDN (2007) Network and system rules of the german transmission system operators, Germany TransmissionCode

[63] Wood AJ, Wollenberg BF (2012) Power generation, operation, and control. Wiley, USA

[64] Wu S, Wu W, Zhang B, Zhang Y (2010) A hybrid dynamic simulation validation strategy by setting V-theta buses with PMU data. Dianli Xitong Zidonghua(Automation of Electric Power Systems) 34(17):12–16

[65] Zhao D, Hu D, He J, Zhang L, Chen N (2016) Model validation of solar PV plant with hybrid data dynamic simulation based on fast-responding generator method. In: MATEC web of conferences, vol 65, p 02006. EDP Sciences

Printed in the United States
By Bookmasters